FACILITATING PEER MENTORING AMONG STEM STUDENTS

This practical guide supports faculty in gaining the skills and expertise they need to successfully facilitate peer mentoring among STEM students. Drawing from several multi-institutional research studies and student feedback, Wendt demonstrates the social, professional, and scholarly benefits of peer mentoring initiatives. Each chapter includes straightforward instruction on how to integrate these practices into mentoring relationships and programs, while also addressing common challenges, providing verbiage and conversation starters, and prompting reflection through smart discussion questions. Easy to read, understand, and implement, this book is an indispensable guide for all faculty within STEM disciplines.

Jillian L. Wendt, EdD, is Professor of Science Education at the University of the District of Columbia, USA.

FACILITATING PEER MENTORING AMONG STEM STUDENTS

Actionable Steps for Faculty

Jillian L. Wendt

NEW YORK AND LONDON

Designed cover image: © Getty Images

First published 2026
by Routledge
605 Third Avenue, New York, NY 10158

and by Routledge
4 Park Square, Milton Park, Abingdon, Oxon, OX14 4RN

Routledge is an imprint of the Taylor & Francis Group, an informa business

© 2026 Jillian L. Wendt

The right of Jillian L. Wendt to be identified as author of this work has been asserted in accordance with sections 77 and 78 of the Copyright, Designs and Patents Act 1988.

All rights reserved. No part of this book may be reprinted or reproduced or utilised in any form or by any electronic, mechanical, or other means, now known or hereafter invented, including photocopying and recording, or in any information storage or retrieval system, without permission in writing from the publishers.

For Product Safety Concerns and Information please contact our EU representative GPSR@taylorandfrancis.com. Taylor & Francis Verlag GmbH, Kaufingerstraße 24, 80331 München, Germany.

Trademark notice: Product or corporate names may be trademarks or registered trademarks, and are used only for identification and explanation without intent to infringe.

ISBN: 978-1-032-91440-4 (hbk)
ISBN: 978-1-032-91439-8 (pbk)
ISBN: 978-1-003-56334-1 (ebk)

DOI: 10.4324/9781003563341

Typeset in Galliard
by KnowledgeWorks Global Ltd.

*This work is dedicated to my students, who have taught me
more about mentoring than any book could
and to my daughter, Madison, who has encouraged me,
motivated me, and inspired me throughout
this entire process.*

CONTENTS

Preface   *x*

PART I
Foundations and Frameworks of Peer Mentoring   **1**

1 Foundations of Mentoring   3

Introduction 3
What Is Mentoring? 5
Why Mentoring? 6
Common Mentoring Structures 9
The Problem 11
The Theoretical and Conceptual Frameworks of
   Mentoring 12
Summary 15
Reflect and Act 16
References 17

2 Understanding Identity   20

Introduction 20
Exploring Identity Theory 20
Summary 32
Reflect and Act 33
References 34

3 Cultural Responsiveness, Motivation,
   and Growth Mindset   36

   Introduction 36
   Cultural Responsiveness 37
   Motivation and the "Push" to Succeed 43
   Attribution Theory of Motivation 44
   Situated Expectancy-Value Theory 45
   Self-Determination Theory 46
   Social-Cognitive Theory 47
   *Tinto's Model of Student Motivation
      and Persistence 48*
   Growth Mindset 48
   Connecting It All 51
   Summary 52
   Reflect and Act 53
   References 54

PART II
Facilitating the Peer Mentoring Relationship   57

4 Getting Started   59

   Introduction 59
   Defining Peer Mentoring 59
   Peer Mentoring Roles 60
   *The Phases of Peer Mentoring: Doing the Actual
      Facilitating 71*
   Facilitating Match and "Fit" 73
   *Facilitating Relationship Establishment
      (Phase 1) 75*
   *Facilitating Purpose, Goals, and Boundaries
      (Phase 2) 75*
   *Supporting the Work of Peer Mentoring
      (Phase 3) 78*
   *Supporting the End of a Relationship
      (Phase 4) 82*
   Summary 83
   Reflect and Act 84
   References 85

5 Leading by Example 87

Introduction 87
Being a Competent Leader 89
Effective Communication 90
Building Trust 93
Understanding Context 94
Boundaries 98
Managing Conflict 100
Ethics in Mentoring 103
Summary 106
Reflect and Act 106
References 108

6 Building Community and Facilitating Opportunity 111

Introduction 111
Community 112
Forging Paths 114
Culture, Norms, and Practices 118
Supporting Creativity 121
Shifts in Relationship Role 122
Summary 123
Reflect and Act 123
References 124

PART III
Resources for Enhancing Knowledge 127

7 Suggested Resources 129

Introduction 129
Additional Suggested Resources 130
Motivation 140
Summary 145
Reflect and Act 145
References 145

Index *149*

PREFACE

The benefits of mentoring programs among racial, ethnic, and gender-diverse students at institutions of higher education have been widely recognized within the research literature. Recent efforts have focused on the impact of mentoring among students interested in matriculating into the science, technology, engineering, and mathematics (STEM) workforce. However, the need for facilitating a robust pipeline of workers to meet the growing needs of the United States' (U.S.) STEM economy has not yet been met—despite years of effort focusing on many different factors. In the post-COVID landscape, especially, our students need targeted support to help them meet their STEM-related career goals. Just as importantly, faculty need resources and skills that will enable them to provide mentorship and guidance to students while simultaneously navigating increased workloads, increasing job insecurity, stagnating salaries, and high rates of burnout.

Peer mentoring has been identified as one solution to this multifaceted problem. In peer mentoring, students mentor each other, reducing the burden on faculty while increasing the academic, social, and psychosocial connections forged between near peers. The exciting part is that recent initiatives have developed and tested rigorous and targeted peer mentoring training for students to help them in gaining the skills and competencies needed to "do" peer mentoring well. These initiatives have resulted in positive outcomes, including higher levels of interest in STEM, higher levels of feelings of self-efficacy, higher levels of motivation, an increased sense of belonging and connection, and even higher levels of persistence among students at various institutions of higher education. Unfortunately, though, tools to assist faculty in learning how to appropriately foster peer mentoring relationships that are effective and efficient are still nearly impossible to find—especially in contexts other than the research laboratory.

In this book, I am focused on attending to this very real and timely faculty need. The overall purpose of the book is to serve as a practical guide for faculty in implementing and facilitating peer mentoring relationships among students—especially those in STEM-related degree programs–within institutions of higher education. Through actionable steps, my hope is that, with this book, faculty will be able to quickly identify the benefits of supporting peer mentoring initiatives and, importantly, gain the necessary skills and expertise to successfully facilitate peer mentoring within their respective institutions. Thus, my aim is to guide faculty (you!) in successfully facilitating peer mentoring relationships among students—increasing faculty members' expertise and skill in mentoring and, therefore, increasing student retention in and completion of STEM degree programs. What follows is based on my own experience in constructing, implementing, testing, and leading peer mentoring programs at various institutions (particularly historically Black colleges or universities, or HBCUs) as well as what is currently available in the research literature.

While this book is primarily written for faculty within STEM programs in higher education, it is widely relevant across various disciplines. So, if you aren't in STEM, that's okay! The key concepts are transferable no matter what particular context you currently find yourself in. I am hopeful that you find this book to be easy to read, understand, and implement, and that you find the questions, reflections, and actionable steps provided helpful in supporting you as you facilitate effective peer mentoring relationships among your students. Happy reading!

PART I
Foundations and Frameworks of Peer Mentoring

1
FOUNDATIONS OF MENTORING

Introduction

Mentoring has become ubiquitous within the field of higher education as well as within the workforce. It has been widely recognized that mentoring can yield positive benefits, from increasing interest in a specific field, to improving rates of retention, self-confidence, and productivity. It is likely that you've had experience as a mentor or mentee at some point during your education or career. What was that experience like? Was it effective or ineffective? What about the experience that was most beneficial to you? Did the mentoring experience improve your academic outcomes, social outcomes, networking opportunities, or other aspects of professional or personal life? Did the experience impact your desire to in turn mentor others?

A multitude of benefits arise from engaging in the mentoring experience for faculty facilitators, mentors, and mentees. Despite these benefits being clear (and, importantly, supported by a wide base of research literature), and despite our inherent desire as educators to nurture and give back, there remains a limit to the amount of time and energy that each of us has to dedicate to mentoring the next generation of learners and workers. It is no secret that educators are often overworked, with class sizes increasing, administrative tasks piling up, and expectations for performance rising while the amount of time available in each day remains the same. This means that we (faculty) collectively must be creative and begin to embrace mentoring structures that capitalize on the resources available to us. Enter peer mentoring.

DOI: 10.4324/9781003563341-2

In this book, the predominant mentoring structure that will be of focus is peer mentoring. This book will be presented through the lens of science, technology, engineering, and mathematics (STEM) faculty at institutions of higher education as this is where the bulk of my own experience rests, although the information in this book will be of benefit to anyone interested in learning more about mentoring. The primary focus of the book will be how faculty can support peer mentoring by serving as knowledgeable guides and facilitators; thus, as *faculty facilitators*. In alignment with contemporary mentoring literature, this book will embrace the cascading mentoring approach, where faculty serve as the most experienced "other" and mentor students (often graduate students) who in turn mentor other less experienced students (often undergraduate students). While the mentoring structure will look different across contexts—and rightly, should—this book will serve as a scaffold for helping faculty to ensure that mentoring can be implemented and facilitated effectively within multiple contexts.

With this in mind, there are a few key concepts that will drive the content of this book. The first concept is that effective mentoring should draw on the experiences and resources of multiple individuals. Information should never flow in one direction or from only one source. Effective mentoring practices are reciprocal practices. The second concept is that, when mentoring is implemented effectively using a faculty-facilitated peer mentoring structure, it can address the lack of available mentors within higher education. This is especially important among populations that have been historically underrepresented (e.g., gender, race, ethnicity). The third concept is that mentoring approaches within which faculty engage must take into consideration the very real potential for burnout. Faculty workloads have increased exponentially, and mentoring endeavors should not add to this workload. When faculty are informed and strategic about mentoring, they can provide the gift of the mentoring opportunity to students in ways that are beneficial to all involved in the relationship without becoming burdened. This book, then, is a guide for faculty wishing to support effective peer mentoring relationships among students in higher education.

While situated within the context of the available research literature and informed by my own experience implementing peer mentoring programs primarily at historically Black institutions, this book is meant to be accessible and easily navigated by faculty in myriad institutions and areas of expertise. With this in mind, I'll do my best to stick to the basics, giving you what you need to facilitate peer mentoring relationships at your institution as well as additional resources for further study should you desire. Think of this book as a user guide rather than a stringent academic text. With that said, I am hopeful that you will find this resource beneficial and, importantly, a tool for supporting students. Let's get started.

What Is Mentoring?

Most of us are familiar with mentoring in one capacity or another, but before we get into the thick of things, I find it important to purposefully define terms. Within the research and practitioner-based literature, multiple definitions of mentoring can be found. The label "mentoring" can indeed serve as an umbrella term for myriad knowledge-providing scenarios. Packard supports this viewpoint, sharing that "mentoring is a broad concept, a term that refers to many different kinds of relationships, programs, and initiatives" (2016, p. 1). McGee and Keller define mentoring as a "reciprocal, dynamic relationship" (2007, p. 316) between two or more individuals. Packard further describes mentoring as "a developmental experience or a type of support intended to advance students toward an important goal" (2016, p. 5). Mentoring is "intended to help students develop, increase their capacity to learn, and encourage persistence in the field" (Packard, 2016, p. 29). Mullen defines mentoring as "a personal or professional relationship between two people–a knowing, experienced professional and a protege or mentee–who commit to an advisory and nonevaluative relationship that often involves a long-term goal" (2005, p. 2).

While no single standard definition of mentoring exists, you likely noticed from the definitions and descriptions I presented that the underlying concepts remain the same. Mentoring is a process in which two or more individuals work together for personal and/or professional development and goal attainment (Rolfe, 2021). It is a unique experience "in that the mentor has no vested interest in outcomes other than positive intention for the mentee" (Rolfe, 2021, p. 12). Likewise, the faculty facilitator wants only positive outcomes for the mentor and mentee. Ultimately, mentoring involves someone serving as a role model for and providing guidance to (Mullen, 2005; Packard, 2016) a differently experienced other or others. For the purposes of this book, I will be adopting Mondisa and Adams's definition of mentoring: "a process in which an experienced individual (a mentor) provides emotional and psychosocial support (e.g., listening, empathizing, offering advice, providing affirmation or an objective perspective), and helps to educate, guide, and counsel a less experienced person" (2022, p. 339).

Most contemporary views of mentoring have embraced a reciprocal approach—one in which all parties engage in both giving and receiving information. (If you recall, this was one of the key concepts that I presented earlier in this chapter.) While it is likely, and quite frankly expected, that one individual has more experience in a certain area than the other, the other individual(s) is likely more experienced in a different area. In other words, everyone in the mentoring relationship has strengths in addition to areas for growth. No single person holds all of the knowledge or experience. This means that everyone in the mentoring relationship has something to give, something to learn, and stands to benefit in some way. Even for the most experienced individual (hello, faculty!), there is always something that can be learned and some sort of knowledge, skill, or strategy that can be shared. Mentoring is a *mutualistic relationship*, to borrow a biological term. That is, mentoring is reciprocal—a form of partnership (Rolfe, 2021)—a partnership that recognizes human possibility (Mullen, 2005) and capitalizes on what each unique person has to offer. It is a collaborative partnership (Ferris & Waldron, 2022). It is a partnership that respects differences among experiences, cultures, and values and embraces the lessons that are learned and knowledge that can be gleaned from those differences. Mentoring is, therefore, an acknowledgement of collective labor (see Figure 1.1).

Why Mentoring?

Most educators are familiar with the term *high-impact practice*—a practice that involves a minimal amount of time and effort yet yields a high return on investment in relation to student outcomes. Mentoring is considered a "high-impact educational practice" (Packard, 2016, p. 5). A rather large body of research literature exists that has demonstrated the many benefits that mentoring can yield across various contexts. For instance, when engaging in mentoring relationships, studies have shown that students can experience an increased interest in their academic work and respective career field, as well as increases in self-efficacy, sense of belonging, persistence, and motivation among other benefits (Burt et al., 2019; Daniels et al., 2019; Jehangir et al., 2022; National Academies of Sciences, Engineering, and Medicine, 2019; Rockinson-Szapkiw, Herring, et al., 2021; Rockinson-Szapkiw, Wendt, & Stephen 2021; Wendt & Jones, 2024a; Wilton et al., 2021). Mentoring affords students the opportunity to make connections that facilitate networking, accessibility to resources, and overall growth in their field of interest. And importantly, this can lead to increased academic success, leading to student retention for institutions (Fehmi et al., 2017), as well as positive career outcomes as students enter the workforce (Saffie-Robertson, 2020).

The benefits of mentoring probably don't surprise you. As each of us is aware, "no one succeeds without assistance" (Fortenberry, 2016, p. xiii). From a developmental perspective, if we consider our lives from infancy on, our growth and development—physical, social, and psychosocial—has been reliant on a "more experienced other" from the moment we entered the world. We relied on the expertise and reciprocity of our caretakers (or in developmental terms, a "skilled helper") in order to meet our basic needs. We relied on the experiences and knowledge that our caretakers had gained

Foundations of Mentoring 7

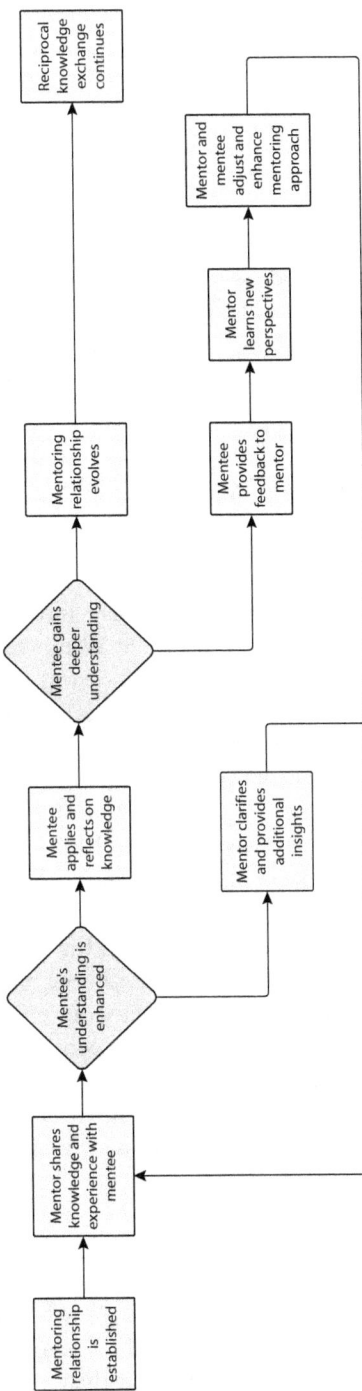

FIGURE 1.1 The mentoring relationship is a reciprocal process between mentor and mentee.

throughout their lives in utilizing language to communicate our needs, feelings, wants, and desires. We relied on the reciprocity of these relationships in many ways—for most of us, a loving give-and-take relationship was a core component of our development of healthy attachment and psychosocial development. As we moved through childhood, our experience increased, yet we continued to be reliant to a varying extent on those around us— peers, teachers, religious leaders, for instance—we continued to gain knowledge, skill, and expertise in navigating the world around us. As time went on, we began to serve as models—whether intentional or not—for those younger and less experienced at life. Humans are social creatures (Vygotsky, 1978, 1986). We grow, develop, and learn within the context of social interactions (Levine & Munsch, 2022). In many ways, then, we've been engaging in mentoring relationships from the beginning of our lives.

The Importance of Intentional Training

While the experiences just described are natural and are a part of human development over time, mentoring relationships for the purpose of enhancing academic, social, and psychosocial skills are most effective when they are *not* left to chance (Rockinson-Szapkiw & Wendt, 2020; Rolfe, 2021). While mentoring in many ways can—and for most intents and purposes should—develop organically, mentoring requires intentional development of skill and knowledge in order for mentoring relationships to be effective, healthy, and meaningful. Mentors and mentees must be trained in effective mentoring practices in order to engage in mentoring relationships that are responsible and impactful (McGee & Keller, 2007; Packard et al., 2014; Pfund et al., 2016; Pon-Barry et al., 2017; Rockinson-Szapkiw, Wendt, & Stephen, 2021; Rockinson-Szapkiw et al., 2020; Wendt & Jones, 2024a). Without intentional training and development, the mentoring relationship can very likely be ineffective and, in some cases, potentially harmful. For instance, without intentional training and development, individuals engaging in mentoring relationships could potentially perpetuate stereotypes and misconceptions or even engage in macro- and micro-aggressions—especially among more vulnerable populations—leading to negative outcomes and causing more harm than good. We don't know what we don't know, right? So intentional training and skills development related to mentoring are absolutely essential.

Experts have also pointed out that mentoring, from a traditional and historical viewpoint, has the tendency to place the more experienced individual in a position of power (Packard, 2016). Oftentimes, those seen as more experienced—those that have been in the field or workforce and have experienced some level of success—forget that their achievements were not obtained without assistance from someone somewhere at some time. As Packard shares, "some individuals who have 'succeeded' seem to have forgotten this objective reality. The tragedy of such forgetfulness is that those in a position of power and influence may not merely decline to provide assistance to others, they can impede the abilities of others to provide such assistance" (2016, p. xiii). Thus, being ill-prepared for engaging in mentoring relationships—especially within the role of faculty facilitator and/or mentor—can easily perpetuate a power differential. A power differential often occurs—and is perpetuated by—the mentor or faculty facilitator having already completed their degree, for instance, either intentionally or unintentionally committing microaggressions or macroaggressions, and/or failing to recognize their own privileges (e.g., race, ethnicity, gender) (Madore & Byrd, 2022). Thus, being aware of the potential for power differential—and actively working to mitigate the negative impacts of such differential—is not only important, but essential. (We'll dive into power dynamics more deeply in Chapter 5.) This reiterates the point that faculty need to be provided with resources and training in how to effectively support mentoring relationships among students as much as students need to be trained in how to be effective mentors and mentees. Faculty facilitation of mentoring and faculty guidance and support should *not* be left to chance. Let's turn our attention now to the different types of mentoring structures so that you can develop a more comprehensive picture of what mentoring might look like.

Common Mentoring Structures

Within the broader umbrella of mentoring, there are multiple types of mentoring relationships. There are formal mentoring relationships, which might historically include some sort of required training, a certain number or type of meetings, and more formalized, stringent expectations. There are informal mentoring relationships that tend to rise more organically and involve meeting, sharing information, or providing guidance more on an as-needed basis. Within these main types of mentoring, there are other nuanced structures. Mentoring might occur in an individualized setting (i.e., one-on-one) or within a group. Mentoring might involve an experienced faculty member guiding a less experienced student (i.e., faculty-student) or follow a cascade approach involving a faculty member, experienced student(s) or research assistant(s), and less experienced student(s) (Packard, 2016). Mentoring might rotate between individuals or even occur between peers or near-peers (i.e., peer mentoring).

How mentoring activities occur can also vary and reside on a continuum from a fully face-to-face mentoring relationship to a fully virtual (i.e., online) mentoring relationship. In the wake of the COVID-19 pandemic, for instance, many institutions were forced into emergency remote instruction, thus facilitating the need for mentoring relationships within the context of higher education to move to the virtual environment. Mentoring truly encompasses a spectrum of structures. There is no one "right" structure for a mentoring relationship as the mentoring relationship should always attend to individual needs and, thus, can take many forms—even changing over the course of the relationship as needs and goals change. Importantly, different approaches are needed for each different mentoring structure and each desired mentoring outcome.

Mentoring can also occur in myriad contexts. Traditional forms of mentoring within the science, technology, engineering, and mathematics (STEM) fields, for instance, tend to occur within the laboratory context, with faculty often serving as the mentor overseeing some course of research investigation while research assistants and/or students carry out the day-to-day activities of the research under the guidance of the mentoring faculty. This is historically where the bulk of research on mentoring within the body of scholarly literature has focused. More recent efforts have explored the benefits of mentoring within the context of STEM, for instance, outside of the research laboratory, attending to mentees' needs above and beyond field-specific skills to overall academic success, social development, psychosocial development, identity development, and sense of belonging, among other constructs (Wendt & Jones, 2024a, 2024b). We'll explore these areas further in subsequent chapters. But for now, it's helpful to understand that the specific context of the mentoring relationship can vary and should ultimately align with the outcomes and goals agreed upon by the mentor(s) and mentee(s).

In this book, the main focus is on peer mentoring among students and how faculty can support students as they engage in this type of relationship. Peer mentoring tends to be a more viable mentoring structure in today's educational environment given the number of students seeking mentors and the lack of overall faculty availability, whether due to an already overloaded schedule, lack of demographic match, or due to sheer volume (Mondisa, 2018). With peer mentoring, students engage in mentoring each other, reducing the burden on faculty while increasing the social and psychosocial connections forged between near-peers.

Peer mentoring can be defined as "a mentoring relationship that occurs between two or more peers, where one peer is usually more experienced than the other(s) and is referred to as the *mentor*. The less experienced peer(s) is referred to as the *mentee*" (Rockinson-Szapkiw et al., 2020 p. 2). Typically, peer mentoring is implemented in an informal manner, although recent initiatives have advocated for the benefit of more formalized peer mentoring (Jones & Wendt, 2025; Wendt & Jones, 2024a, 2024b). The benefits of peer mentoring are many, but the most immediate is that diverse faculty that mirror the demographics of the student population (as well as others that find themselves in the mentee role) are limited (Ferris & Waldron, 2022), and, simultaneously, mentoring requires time and resources that, quite frankly, most faculty do not have at their disposal. With peer mentoring, students work together to facilitate a mentoring relationship and, when done well, reap the benefits of near-peers as well as those of a guiding faculty member or *faculty facilitator*.

The Problem

While mentoring is widely accepted as a beneficial practice, it isn't always implemented well. We know, for instance, that mentoring—especially peer mentoring—is more effective when mentors and mentees are provided the training, resources, and supports that enable them to engage in the mentoring relationship from a place of knowledge and confidence. Recent efforts have attended to this need and have demonstrated overall success among students (Rockinson-Szapkiw, Herring, et al., 2021; Rockinson-Szapkiw, Wendt, & Stephen, 2021; Wendt & Jones, 2024a, 2024b).

But what about faculty? Studies have supported that, even when engaging in peer mentoring relationships, mentors and mentees want faculty guidance (Wendt & Jones, 2024a, 2024b). That is, students recognize and respect the expertise, skills, and networks that faculty have developed throughout their careers. In fact, in my own studies on peer mentoring, student participants have reported wanting increased interaction from faculty while simultaneously reaping the benefits of near-peer relationships (Jones & Wendt, 2025; Wendt & Jones, 2024a, 2024b). Research overall supports that "a network-based approach to mentoring" (Packard, 2016, p. 31) enables individuals with varied skill sets, different interests, and in different life and/or career stages to engage in effectively practicing reciprocity. This is because "a diverse network of mentors is more likely to produce more and better quality mentoring than relying on just one mentor to do it all" (Packard, 2016, p. 31). Remember earlier in the chapter when I reiterated that no one person holds all of the knowledge? That we all have something to learn as well as something to give? Faculty, therefore, are necessary within mentoring relationships—even peer and near-peer mentoring relationships that occur primarily among students.

With this in mind, should it be assumed that faculty, given their extensive experience in their respective fields, also have experience in how to most effectively mentor (or guide effective mentoring relationships among) students? Or should faculty also be provided with resources, supports, and training in evidence-based practices related to mentoring to ensure effective and healthy mentoring relationships? The latter is my belief, supported by my experience in collaboratively creating, implementing, leading, and researching mentoring among students at various institutions—particularly those that have been historically underserved and underrepresented (especially in STEM fields) and, thus, stand to benefit the most from mentoring (Fortenberry, 2016; McGee, 2016).

The Theoretical and Conceptual Frameworks of Mentoring

Mentoring is grounded within multiple theoretical and conceptual frameworks, and it is important to have a basic understanding of these frameworks in order to understand how to effectively support peer-mentoring relationships. For the purposes of brevity, we'll focus on only a few of these frameworks within this book, giving precedence to those that are most common within the current research literature and those that will assist you as faculty in best facilitating students' engagement in peer mentoring relationships. Let's begin with the Institutional Departure Model (Tinto, 1975, 1988, 1993).

Institutional Departure Model

The Institutional Departure Model (Tinto, 1975, 1988, 1993) relies on the core understanding that every individual has their own unique personal attributes, familial background, and prior experiences, and is a model that is often aligned with mentoring initiatives within the body of mentoring literature. For clarification purposes, let me first define the major constructs associated with the model. *Personal attributes* may include race, ethnicity, culture, and gender. *Familial backgrounds* include the individual's familial socioeconomic status as well as level(s) of their parents' education. And *prior experiences* include both academic experiences and socioemotional experiences. When entering an institution (in our case, an institution of higher education), these attributes will influence the individual's ability to integrate within the culture of that particular institution. Those individuals that are able to integrate well into the institutional culture and climate will be more apt to experience success. Those individuals that are unable to integrate as well will be more likely to experience challenges in areas such as commitment, persistence, sense of community, and sense of belonging (Tinto, 1975). Thus, it is the degree to which the individual can integrate that, in many ways, determines one's academic and social outcomes, including academic and social successes.

Mentoring, in alignment with the Institutional Departure Model, has demonstrated benefits to students' academic and social outcomes by promoting integration. This is because mentoring assists with the development of a sense of belonging as well as an overall sense of community by providing such things as social interaction, networking, and access to resources that otherwise may not be accessible (Wilton et al., 2021). Peer mentoring is especially beneficial to supporting integration, as the reciprocal nature of engaging in relationships with near peers provides both the peer mentors and peer mentees with psychosocial support, instrumental support, and academic support. Psychosocial support is provided through the adoption of feelings of enhanced competence. Instrumental support is provided by nature of engagement with near peers, developing a sense of belonging within a particular space, and assisting with goal development to meet an individualized—yet often shared—goal. Academic support is provided through access to the development of content-area knowledge as well as the development of more nuanced career skills. Just as importantly, integration supports students' persistence in their degree programs (Wilton et al., 2021) and retention within the institution (Tinto, 1987). Thus, by supporting students' integration into the institution, the department, and their specific field, mentoring serves as one factor in ensuring students' success—not only within their degree programs, but within the future workforce.

Social Cognitive Career Theory

The next framework that I want to share with you is Social Cognitive Career Theory (SCCT) (Lent et al., 1994). SCCT is a theoretical framework which is also closely aligned with mentoring within the research literature. SCCT relies on the value that an individual places within a particular discipline. When an individual places high value within the discipline, their likelihood to persist, their motivation, and their levels of engagement are enhanced. If the individual, however, places low value in a discipline, their interest is in turn lower and, thus, they are less likely to persist, experience lower levels of motivation, and are less apt to engage in opportunities within the discipline. Persistence, motivation, and engagement are each important as they facilitate opportunities for growth—in trying new tasks, developing mastery, and obtaining useful skills (Bandura, 1977, 1997). When an individual experiences growth, they also experience enhanced levels of self-efficacy and a sense of belonging. Effective mentoring relationships can facilitate students' growth and, therefore, the value that they place on their courses, degree, and overall field of study.

Self-Efficacy Theory

The next theory that is commonly found within the mentoring literature is Self-Efficacy Theory (Bandura, 1977). Self-efficacy is defined as one's belief in their own ability to complete a task successfully (Bandura, 1977).

When students have high levels of self-efficacy, they are more likely to retain interest in their field of study, persist, and, importantly, experience success (Fouad et al., 2016). Simply put, there are four areas that influence students' levels of self-efficacy: performance accomplishment, vicarious experience, social persuasion, and physiological response (Bandura, 1977, 2006). Let me define each. *Performance accomplishment* is experienced when a student successfully completes a given task. *Vicarious experience* occurs when a student observes someone else successfully complete a given task. *Social persuasion* occurs when a student receives from others, either encouragement or discouragement related to their ability to engage in a task successfully. And, finally, *physiological response* occurs as a result of a student's reaction to a specific experience. Together, these four sources either support or inhibit the development of self-efficacy. In effective mentoring relationships, each of the four sources—and, in turn, self-efficacy—is supported.

Identity Theory

Identity Theory is the final area that I want to highlight for now, and it ties in nicely with Self-Efficacy Theory. Self-efficacy assists in the development of identity (Chemers et al., 2011), especially as it relates to professional identity. Identity is typically grounded within Erikson's theory of psychosocial development (Erikson, 1950) and can be defined as "an individual's sense of self and connections built within their environment" (Ivey & Parish, 2021, p. 155). The development of identity, as well as the recognition of how identities intersect, is important in students' development of a sense of belonging, confidence, competence, and motivation within a particular field of study and career field (Carlone, 2007; Clark et al., 2016; Hill et al., 2010; Wendt & Jones, 2024a, 2024b). Mentoring can support students in both recognizing their identity(ies), reconciling intersecting identities, and determining if their identities "fit" within their respective fields. While the concept of identity is not new, only recently has research really begun to explore the importance of students' development of identity within mentoring contexts, as well as the importance of identity in supporting students' academic and career success. I'll take a deeper dive into identity theory in Chapter 2.

While the overview provided here is certainly not a comprehensive list of theoretical and conceptual frameworks among which mentoring can be grounded, I'm hopeful that it has provided you with a little bit of background so that you can more fully understand *how* mentoring can lead to positive student outcomes. Table 1.1 provides a very condensed version,

TABLE 1.1 Key tenets of commonly encountered theories supporting the benefits of mentoring

Institutional Departure Model	(Tinto, 1975, 1988, 1993) • Personal attributes, familial backgrounds, and prior experiences Influence students' ability to integrate within the institution. • Mentoring can promote integration by developing a sense of belonging and sense of community by providing Interactions, networking, and access to resources.
Social Cognitive Career Theory	(Lent et al., 1994) • Success relies on the value that students place on a specific discipline. • Increased value leads to an increased likelihood of persistence, increased motivation, and increased levels of engagement. • Mentoring can facilitate experiences for growth, which can lead to increased self-efficacy and increased sense of belonging, which in turn can lead to an increased value on a specific discipline.
Self-Efficacy Theory	(Bandura, 1977) • Posits that a student's belief in their own ability to complete a task successfully (self-efficacy) is influenced by performance accomplishment, vicarious experience, social persuasion, and physiological response. • Mentoring supports performance accomplishment, vicarious experience, social persuasion, and physiological response, which in turn supports the development of self-efficacy.
Identity Theory	(Erikson, 1950) • An individual's sense of self is influenced by connections that are built within their environment. • Mentoring supports identity recognition, reconciliation of intersecting identities, and determination of "fit" within a specific disciplinary field, social space, or physical space.

and you are encouraged to seek out additional resources (see Chapter 7) if you want to know more.

Summary

In this chapter, we have defined mentoring, including the focus of this book—peer mentoring through the lens of STEM. The purpose of the book—serving as a resource for faculty interested in implementing and supporting peer mentoring relationships among students—was described. The most common mentoring structures were outlined, as were the most prevalent theoretical and conceptual frameworks upon which mentoring is based. As we wrap up this chapter and move forward in our learning, take a few moments to consider the questions and prompts presented in the Reflect and Act section.

Reflect and Act

1 What do you feel are the benefits of peer mentoring for faculty? For peer mentors? For peer mentees?
2 Why is it important for faculty to be provided with resources and opportunities for skills development prior to facilitating peer mentoring relationships?
3 Reflect on your experiences in mentoring by considering the following questions. With which forms of mentoring, if any, have you had experience as a mentor? As a mentee? As a faculty facilitator? Was the particular form or structure of mentoring appropriate for your goals (if you were a mentor or mentee), your mentees' goals (if you were a mentor), or your students' goals (if you were a faculty facilitator)? What about the mentoring structure worked? What about the mentoring structure did not work? What would you do differently if engaging in a mentoring experience in the future, particularly as a faculty facilitator within a peer mentoring relationship? You might find it helpful to journal about these experiences so that you can refer back to them as you continue through this book.
4 Take some time to reflect on the messages that you have received from your students in particular. What feedback, whether direct or indirect, have you received from students about the types of resources, supports, or experiences that they believe they need in order to be successful? This is particularly important if the feedback was provided within the context of a mentoring relationship. Considering this feedback, what form or forms of mentoring would likely best meet your students' needs moving forward?
5 Think about the major theoretical and conceptual frameworks found within the mentoring literature as presented in this chapter. How might the various theoretical and conceptual frameworks shared in this chapter align with your observations of mentoring relationships? Which framework is most closely aligned with your worldview and philosophy of teaching? How might this particular framework inform your practices as a faculty facilitator within a peer mentoring relationship?

References

Bandura, A. (1977). Self-efficacy: Toward a unifying theory of behavioral change. *Psychological Review*, *84*(2), 191–215.
Bandura, A. (1997). *Self-efficacy: The exercise of control*. W. H. Freeman and Company.
Bandura, A. (2006). Toward a psychology of human agency. *Perspectives on Psychological Science*, *1*(2), 164–180. https://www.jstor.org/stable/40212163
Burt, B. A., Williams, K. L., & Palmer, G. J. M. (2019). It takes a village: The role of emic and etic adaptive strengths in the persistence of Black men in engineering graduate programs. *American Educational Research Journal*, *56*(1), 39–74.
Carlone, H. B. J. A. (2007). Understanding the science experiences of successful women of color: Science identity as an analytic lens. *Journal of Research in Science Teaching*, *44*(8), 1187–1218.
Chemers, M. M., Zurbriggen, E., Syed, M., Goza, B. K., & Bearman, S. (2011). The role of efficacy and identity in science career commitment among underrepresented minority students. *Journal of Social Issues*, *67*(3), 469–491.
Clark, S. L., Dyar, C., Maung, N., & London, B. (2016). Psychosocial pathways to STEM engagement among graduate students in the life sciences. *CBE-Life Sciences Education*, *18*, 1–13.
Daniels, H. A., Grineski, S. E., Collins, T. W., & Frederick, A. H. (2019). Navigating social relationships with mentors and peers: Comfort and belonging among men and women in STEM summer research programs. *CBE-Life Sciences Education*, *18*. https://doi.org/10.1187/cbe.18-08-0150
Erikson, E. H. (1950). *Childhood and society*. Norton.
Fehmi, D., Braun, T. F., & Gublo, K. (2017). Peer mentor program for the general chemistry laboratory designed to improve undergraduate STEM retention. *Journal of Chemical Education*, *94*(12), 1873–1880.
Ferris, S. P., & Waldron, K. (2022). Learning from senior women leaders: In their own words. In B. Cozza & C. Parnther (Eds.), *Voices from women leaders on success in higher education: Pipelines, pathways, and promotion* (pp. 81–82). Routledge.
Fortenberry, N. L. (2016). Foreword. In B. W. Packard (Ed.), *Successful STEM mentoring initiatives for underrepresented students: A research-based guide for faculty and administrators* (pp. xiii–xiv). Stylus Publishing, Inc.
Fouad, N., Singh, R., Cappaert, K., Chang, W., & Wan, M. (2016). Comparison of women engineers who persist in or depart from engineering. *Journal of Vocational Behavior*, *92*, 79–93.
Hill, C., Corbett, C., & St. Rose, A. (2010). *Why so few? Women in science, technology, engineering, and mathematics*. Association for the Advancement of University Women.
Ivey, S. S., & Parish, S. G. (2021). Cultivating STEM identity through the peer mentoring relationship. In A. J. Rockinson-Szapkiw, J. L. Wendt, & K. Wade-Jaimes (Eds.), *Navigating the peer mentoring relationship: A handbook for women and other underrepresented populations in STEM* (pp. 153–165). Kendall Hunt Publishing Company.
Jehangir, R. R., Stebleton, M. J., & Collins, K. (2022). STEM stories: Fostering STEM persistence for underrepresented minority students attending predominantly white institutions. *Journal of Career Development*, *50*(1), 87–103.
Jones, V. O., & Wendt, J. L. (2025). Encouraging confidence: The impact of an online peer mentoring program on women peer mentees in STEM at two HBCUs. *Trends in Higher Education*, *4*(3). https://doi.org/10.3390/higheredu4010003

Lent, R., Brown, S., & Hackett, G. (1994). Toward a unifying social cognitive theory of career and academic interest, choice, and performance. *Journal of Vocational Behavior, 45*(1), 79–122.

Levine, L. E., & Munsch, J. (2022). *Child development: An active learning approach* (4th ed.). Sage.

Madore, M. R., & Byrd, D. (2022). Optimizing mentoring relationships with persons from historically marginalized communities through the communication of difficult dialogues. *Journal of Clinical and Experimental Neuropsychology, 44*(5–6), 441–449.

McGee, E. O. (2016). Devalued Black and Latino racial identities. A by-product of STEM college culture? *American Educational Research Journal, 53*(6), 1626–1662. https://doi.org/10.3102/0002831216676572

McGee, E. O., & Keller, J. L. (2007). Identifying future scientists: Predicting persistence into research training. *Life Science Education, 6*(4), 316–331.

Mondisa, J. (2018). Examining the mentoring approaches of African-American mentors. *Journal of African American Studies, 22*, 293–308.

Mondisa, J., & Adams, R. S. (2022). A learning partnerships perspective of how mentors help proteges develop self-authorship. *Journal of Diversity in Higher Education, 15*(3), 337–353.

Mullen, C. A. (2005). *Mentorship*. Peter Lang.

National Academies of Sciences, Engineering, and Medicine. (2019). *The science of effective mentorship in STEMM*. The National Academies Press. https://doi.org/10.17226/25568

Packard, B. W. (2016). *Successful STEM mentoring initiatives for underrepresented students: A research-based guide for faculty and administrators*. Stylus Publishing, LLC.

Packard, B. W., Marciano, V. N., Payne, J. M., Bledzki, L. A., & Woodard, C. T. (2014). Negotiating peer mentoring roles in undergraduate research lab settings. *Mentoring & Tutoring: Partnership in Learning, 22*(5), 433–445.

Pfund, C., Byars-Winston, A., Branchaw, J., Hurtado, S., & Eagan, K. (2016). Defining attributes and metrics of effective research mentoring relationships. *AIDS and Behavior, 20*, 238–248.

Pon-Barry, H., Packard, B. W., & St. John, A. (2017). Expanding capacity and promoting inclusion in introductory computer science: A focus on near-peer mentor preparation and code review. *Computer Science Education, 27*(1), 54–77.

Rockinson-Szapkiw, A., Herring, J., Gishbaugher, J., & Wendt, J. L. (2021). A case for a virtual STEM peer mentoring experience: The experience of racial and ethnic minority women mentees. *International Journal of Mentoring and Coaching in Education, 10*(3), 267–283. https://doi.org/10.1108/IJMCE-08-2020-0053

Rockinson-Szapkiw, A., & Wendt, J. L. (2020). The benefits and challenges of a blended peer mentoring program for women peer mentors in STEM. *International Journal on Mentoring and Coaching in Education, 10*(1), 1–16.

Rockinson-Szapkiw, A., Wendt, J. L., & Stephen, J. S. (2021). The efficacy of a blended peer mentoring experience for racial and ethnic minority women in STEM pilot study: Academic, professional, and psychosocial outcomes for mentors and mentees. *Journal for STEM Education Research, 4*, 173–193. https://doi.org/10.1007/s41979-020-00048-6

Rockinson-Szapkiw, A., Wendt, J. L., & Wade-Jaimes, K. (2020). *Navigating the peer mentoring relationship: A handbook for women and other underrepresented populations in STEM*. Kendall Hunt Publishing Company.

Rolfe, A. (2021). *Mentoring mindset, skills and tools*. Mentoring Works.

Saffie-Robertson, M. C. (2020). It's not you, it's me: An exploration of mentoring experiences for women in STEM. *Sex Roles, 83*, 566–579.

Tinto, V. (1975). Dropout from higher education: A theoretical synthesis of recent research. *Review of Educational Research*, *45*(1), 89–125.
Tinto, V. (1987). *Leaving college: Rethinking the causes and cures of student attrition*. University of Chicago Press.
Tinto, V. (1988). Stages of student departure from institutions of higher education. In V. Tinto (Ed.), *Leaving college: Rethinking the causes and cures of student attrition* (pp. 84–137). University of Chicago Press.
Tinto, V. (1993). *Leaving college: Rethinking the causes and cures of student attrition* (2nd ed.). University of Chicago Press.
Vygotsky, L. (1978). *Mind in society: The development of higher psychological processes*. Harvard University Press.
Vygotsky, L. (1986). *Thought and language*. Harvard University Press.
Wendt, J. L., & Jones, V. O. (2024a). Peer mentors' experiences in an online STEM peer mentoring program: "Beacons of light." *International Journal of Mentoring and Coaching in Education*, *13*(3). https://doi.org/10.1108/IJMCE-03-2023-0033
Wendt, J. L., & Jones, V. O. (2024b). Supporting BIPOC males in STEM: Insights from a case study on online peer mentoring. *Journal of Research in STEM Education*, *10*(1–2), 89–113. https://doi.org/10.51355/j-stem.2024.145
Wilton, M., Katz, D., Claremont, A., Gonzalez-Nino, E., Foltz, K. R., & Christoffersen, R. E. (2021). Improving academic performance and retention of first-year biology students through a scalable peer mentorship program. *CBE-Life Sciences Education*, *20*, 1–13.

2
UNDERSTANDING IDENTITY

Introduction

In Chapter 1, I provided a brief overview of the most relevant conceptual and theoretical frameworks among which mentoring experiences are situated–one of which was Identity Theory. In this chapter, I'm going to dive a bit deeper into Identity Theory, especially since research is only beginning to explore the impact of mentoring on the development of students' identity—particularly within the context of science, technology, engineering, and mathematics (STEM). I'll also share what is meant by intersecting identities as well as steps for reflecting on your own identity(ies) so that you can more effectively support your students. While many of the concepts in this chapter will be applicable across mentoring contexts, keep in mind that I will be focusing mainly on identity within the context of faculty facilitating students' peer mentoring relationships. Let's get started.

Exploring Identity Theory

Identity Theory is grounded within Erikson's (1950) theory of psychosocial development. Identity can be defined as "an individual's sense of self and connections built within their environment" (Ivey & Parish, 2021, p. 155). Identity is influenced by a student's personal perceptions of what they can do well (strengths) as well as areas of growth (weaknesses). It is ultimately influenced by many factors and, likewise, can contain multiple layers. For instance, identity can be influenced by "gender identity, religion, ethnicity and race, economic class/socioeconomic status, name/family, age, place (geography, national territory), perception of belonging, language, [and] exceptionality (whether gifted or challenged)" (Miller, 2021, p. 3). In extremely simple terms, identity can be thought of as how the person sees themself within the world–in particular, within which group they feel they belong–and, thus, will be influenced by the entirety of their experiences and perceptions.

Identity is important as it situates a person within a specific context, with a specific set of characteristics, and a specific role or place within the world. As Atkins and colleagues (2020, p. 2) share, "when individuals claim an identity, they attribute themselves a set of meanings about their role, group membership, or unique personhood. These meanings are then communicated to others through behaviors and social interactions," and, through the reactions of others, are either affirmed or challenged. When an individual's identity is accepted, respected, and even fostered, their identity is affirmed. In other words, their affinity for their identity—their belief in who they are as a person—is reinforced. This reinforcement may result in an increased level of comfort with who they are within the world. On the other hand, when an individual's identity is rejected, disrespected, or refuted, their identity is challenged. This challenge may result in distress as the individual may not be easily able to reconcile the differences between how others perceive them and how they perceive themselves within the world.

As identity is multifaceted, a person will likely have more than one identity. These multiple identities shape who they are and how they see themselves as fitting (or not fitting) within the world. For example, it is unlikely that a person would identify based on gender alone (i.e., woman). However, it is likely that a person would identify, for example, based on gender, race, familial role, and perhaps profession (i.e., woman, White, mother, educator). These multiple identities overlap and, in so doing, intersect. What this means is that a person might perceive their identities as both separate and overlapping and, in many cases, influencing each other—thus, intersecting. Identities are complex!

Let me give you an example by sharing a little about my identities. I personally can identify as an educator and mother. I could identify as an educator and a mother separately, but in reality, these two identities overlap and intersect. It would be difficult for me to view my role as a mother completely apart and separate from my role as an educator given the constraints that academia (not unlike other professions) presents to motherhood. And the lessons that I have learned through my experience being a mother have informed my knowledge and practices as an educator, especially as it relates to applying theories (developmental theory, for instance) and seeing how theories play out in real-time. My understanding of child development, as one example, has greatly increased as a result of being a mother and watching my daughter journey through different developmental phases.

This enhanced understanding has greatly influenced how I go about my work as an educator. With all of this in mind, I might more accurately identify as an academic mother–honoring the fact that one identity influences and is in turn influenced by the other.

It is important to remember, though, that identity is fluid–it is both socially constructed as well as personally constructed (Miller, 2021). Identity can and likely will change over time as the meanings associated with one's identity are supported or challenged within the context of social interactions. Identities that are reinforced or affirmed are likely to persist over longer periods of time (Atkins et al., 2020), whereas those that are challenged or refuted are less likely to persist. Further, new experiences may facilitate revisions and modifications of one's identity. These new experiences might influence an individual to alter their existing identity or to adopt a completely different identity. Remember, too, that the overlapping and intersecting of identities (intersectionality) can also change, leading to innumerable combinations of identities.

While we are on the topic of intersectionality, please do not be misled into believing that intersectionality is the same as diversity. Intersectionality is, in fact, *not* the same as diversity (Love, 2019). Diversity is often thought of in simplistic terms based on singular identities. However, intersectionality introduces myriad combinations of identities that individually and collectively "reinforce each other to create new categories" (Taylor, 2017, p. 4). As Love (2019) shares, "'Intersectionality' is more than counting representation in a room or within a group; it is understanding community power, or its lack, and ensuring inclusivity in social justice movements" (p. 3). It is, thus, recognizing the individual as a whole person—a complex, multifaceted, unique human being.

Let me give you another example. As I mentioned previously, my own identities are complex. For instance, I identify as a White woman, a mother, an academic, an avid reader, and a writer. As shown in Figure 2.1, these multiple identities that I hold overlap in many ways. For instance, I cannot identify as a mother without also considering my identity as an academic–they are not exclusive from one another, and my identity as one is impacted by and further impacts the other. I cannot identify as a reader without also considering my identity as a writer–one influences and is in turn influenced by another. While intersecting identities may or may not be as easily reconciled as they are in my particular example (for instance, consider how a gender diverse individual may have difficulty reconciling an intersecting identity as an adherent to a strict religious order that views gender as only binary), the example shown in Figure 2.1 can serve as one example of the multiple facets–as well as the intersection–of an individual's identity in very simplified terms.

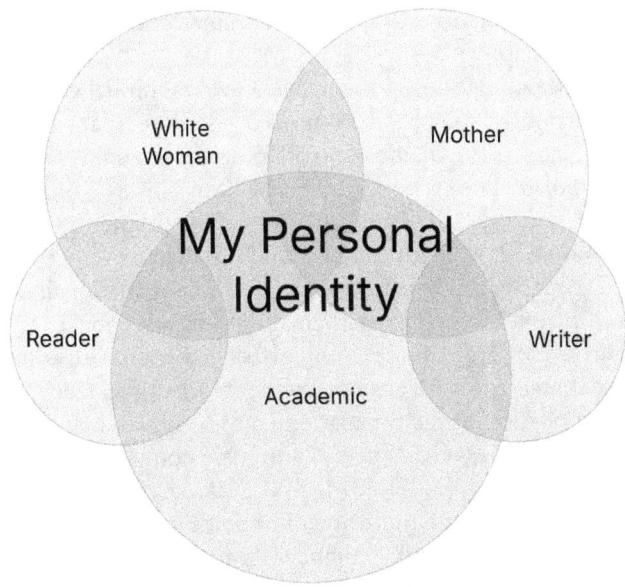

FIGURE 2.1 A representation of selected personal identities of the author.

Why Is Identity Important?

Identity is important as it is the way in which an individual perceives that they belong within a particular group, the meaning that is associated with their level of belonging, and a representation of their unique personhood. It influences their perceived place in the world personally, socially, and professionally. Ultimately, it determines whether an individual feels that they "fit" within a particular group, space, or field. Thus, it is closely linked to a sense of belonging. Identity is also closely linked to an individual's beliefs in their own ability, competence, likelihood of success, and belonging particular to their field of study (Moore et al., 2018); thus, their development of an academic identity.

Let's consider, in particular, identity within STEM fields as one specific example. Historically, women, non-binary, and racially and ethnically marginalized individuals have been underrepresented within STEM degree programs and career fields and, to date, *continue* to be underrepresented in most STEM fields (National Science Foundation [NSF], 2023, 2024). While there are many reasons for this, overall, women, non-binary, and racially and ethnically marginalized populations have largely been seen as not "fitting" within STEM fields (for the purposes of brevity, I'll refer to these particular populations as simply "marginalized" throughout the remainder of this chapter).

Marginalized populations, for instance, have traditionally been relegated to unpaid or poorly paid roles compared to those that encompass the majority population (the majority population being, simply, White males). This has historically been true across fields. Typically, STEM fields tend to pay more, be more highly respected, and carry a higher level of prestige than many non-STEM fields (understanding that, of course, there will always be exceptions). The rates of acceptance of marginalized populations into STEM fields such as mathematics, research science, engineering, and other areas that are considered highly skilled and more complex fields remain far below those who are not considered marginalized. Thus, there is a significant portion of the population (those who are marginalized) that is overall being paid less than their White male counterparts for engaging in the *same* STEM work (NSF, 2024; SWE, 2023). (For the record, let me also acknowledge that yes—there are a few fields in which women are paid more than men. Environmental engineering per the SWE (2023) report is one. My point is that disparity exists.) Note that this is not an issue that is unique to the United States. A recent study has demonstrated that this is a pervasive and global issue (Zajac et al., 2025).

In considering STEM fields in particular, those that are marginalized that have braved entrance to these fields—whether through degree programs or career choice—have often been met with an unwelcoming, chilly climate (Wilton et al., 2021)—one that does not perceive them as belonging within that particular space and that may even intentionally be hostile to the point of pushing them out of the space. This could be for many reasons, such as stereotypes that have situated them as less intelligent, less capable, having less flexibility (consider the traditional roles of women as mothers, for example), or simply being perceived as unable to meet the rigorous demands of "prestigious" fields.

Research has consistently documented the chilly climate that marginalized individuals can experience within STEM (Clark et al., 2016; Daniels et al., 2019; Hurtado et al., 2007) as well as the potential negative impacts on the persistence and retention of a large swath of the talent pool as a result (Huderson & Huderson, 2019; Jones & Wendt, 2025; Wendt & Jones, 2024a). Let's take women, for instance—especially those that are mothers. A traditional view holds that women—in particular, mothers—belong at home. Their role is to be a good wife, oversee the daily functioning of the home, and raise the children. Despite myriad changes in this (outdated) societal perspective, there are some that still hold tight to this view. (If you don't believe me, check out the recent Trad Wife trend (Elmhirst, 2024).) Because of this deeply rooted view, in some STEM environments, women are sometimes perceived as being outsiders that do not belong. They are not embraced as being a vital part of the workforce. They are not mentored and supported to the extent that their male counterparts may be. They are given more menial, often secretarial tasks (note-taker, anyone?). They are not included in the same manner as their male counterparts. This has been widely documented within the research literature and persists in many STEM environments today. (Note that I also recognize that this phenomenon extends to other marginalized populations as well—not just women. And there is a difference between those that actively *choose* to embrace more traditional gender roles and those who have traditional gender roles imposed on them involuntarily. This is a very important distinction.)

Looking at the research literature, many women have reported that they have experienced hostility within the STEM environment–within their courses, degree programs, and places of work (Clark et al., 2016). They report perceiving that the value that they bring to the STEM space is often overlooked and not seen as significant or important. They are often seen as "less naturally gifted [or] skilled" (Clark et al., 2016, p. 2) than their male counterparts. By simply identifying as a woman, these individuals are perceived (and thus, begin to perceive themselves) as not belonging and not fitting in; that, despite their commitment, talent, and hard work, they will not be perceived as valued to the same extent to which their male counterparts will be. Thus, women may decide either not to pursue STEM degrees or careers, and if they do, are likely to leave the field prematurely. Their identity as a STEM professional, for instance, is challenged and refuted. As Clark and colleagues (2016, p. 3) share, when a person's identity conflicts with the space that they are in, "the result may be anxiety and stress that leads an individual to try to reduce that tension or dissonance." In other words, the individual disengages and leaves when they perceive that their identity is not congruent with the environment.

Ultimately, in addition to the ridiculousness of these stereotypes and how inaccurate these perceptions might be, the lack of a welcoming climate for *all* individuals has resulted in a large portion of the population's knowledge, experience, and talent being ignored and, in some cases, lost completely (Clark et al., 2016; Jones & Wendt, 2025; Saffie-Robertson, 2020; Wendt & Jones, 2024a; Wendt & Jones, 2024b). "All individuals adopt multiple layers of personal and social identities ..., and those identities become a lens through which individuals view the world, the decisions they make, and the sense of connection they have with others" (Clark et al., 2016, p. 3). Thus, it is essential that space is made, especially within STEM, for individuals of differing identities; thus, capitalizing on the unique insights, experiences, talents, and gifts that these individuals bring to the field rather than practicing exclusion that ultimately does harm to us all. Part of your role as a faculty facilitator is to ensure a welcoming and safe space for individuals with differing identities—no matter what those identities may be.

The Impact of Mentoring on Identity

The good news is that mentoring has been shown to support the development of identity—especially among marginalized populations (Garcia-Murillo et al., 2023; Murrell et al., 2021); thus, supporting the participation of diverse populations within fields that, quite frankly, could benefit from this "lost" talent (Clark et al., 2016; Saffie-Robertson, 2020). Mentoring has the capacity to support a sense of belonging, increase levels of motivation and persistence, and foster self-efficacy (Jones & Wendt, 2025; Rockinson-Szapkiw & Wendt, 2020; Rockinson-Szapkiw et al., 2022; Wendt & Jones, 2024b). I personally have seen this firsthand in my own research on peer mentoring among marginalized populations at historically Black colleges and universities (HBCUs)!

Mentoring, in short, can support identity development as well as identity compatibility. Increased identity compatibility leads to increased well-being and enhanced performance (Clark et al., 2016). Identity compatibility correlates directly to goals, progress toward meeting goals, and the decisions that are made to either remain within the field or not. Identity compatibility may determine whether an individual experiences "confidence in their academic skills and social belonging" within their field (Clark et al., 2016, p. 3). When an individual is mentored by someone who not only honors their identity but also ensures a safe and welcoming environment in recognition of their identity, positive outcomes can result (Garcia-Murillo et al., 2023; Murrell et al., 2021). Again, understanding this point is imperative for faculty facilitators.

Interrogating Your Own Identity

As a faculty facilitator, how can you most effectively recognize and honor a person's identity, though? The first step is to interrogate your own identity. That is, you must take the time to reflect on how you identify, where your identities may or may not overlap, and the extent to which your identities are compatible with your own goals. What follows are suggested steps for you to take to interrogate your own identity based on a workshop presented and published by O'Sullivan and Irby (2021). The following steps are best completed with others. Being able to share with other individuals, discuss, and reflect is a key component to identifying and more fully understanding your own identity.

Introductions

First, find one to three people that you can provide an introduction of yourself to. These could be people that know you or people that don't know you. Given the context of faculty identity development, these people would be best situated within your institutional academic environment. Explain to them that you are doing an exercise on identity development to help you better connect with and support students. Then, introduce yourself to them. Share whatever you feel best describes you, but keep it brief (think 1–2 minutes at most). If it helps, think of this as being a sort of *elevator speech* on who you are.

Reflection

After introducing yourself to each person, write down what you have said to each person. This does not need to be verbatim, but should capture as much as possible about the nuances of what you said and how. After writing down what you said, reflect on the personal and professional roles that you shared. What roles did you share? What details did you provide about the meaning of those roles to you? Did you share anything about organizational affiliations? What about these organizational affiliations appeals to you? How do these affiliations align with your values, goals, and way of seeing yourself and others within the world? In a different setting, perhaps outside of the institution, would you have said anything differently? If so, what and why? Take brief notes on your reflection as you have now articulated your identities.

Identify Identities

Identities can be organized into three main categories: personal identities, professional identities, and organizational identities (O'Sullivan & Irby, 2021). Table 2.1 provides an example of how identities can be categorized.

TABLE 2.1 The three major categories into which identities can be categorized

Personal Identities	*Professional Identities*	*Organizational Identities*
Examples: Mother, father, daughter, son, reader, crafter, cyclist	**Examples:** Academic, engineer, biologist, director, investigator, researcher	**Examples:** College/university professor, department chair, sports fan, club member, religious affiliation, political affiliation

Identity Theory assists in describing the ways in which individuals view themselves as well as how they believe that others view them within a particular space or within the world (Roccas & Brewer, 2002). Within this framework, an individual will then determine the extent to which their identities are similar to others', indicating fit or match, or the extent to which their identities are different from others'. From the notes that you have taken thus far, clearly define—that is, write down—your identities and organize them into each of the three main categories.

Categorize and Compartmentalize

Identities can be categorized as well as compartmentalized (Roccas & Brewer, 2002). An individual might have one single identity. An individual might have several identities, each of which is distinct and separate, as I have shared earlier. An individual might have several identities, some of which overlap or merge. Look at the identity(ies) that you have written down. Do you have only one singular identity? If you wrote down multiple identities, do these identities exist separately, distinct, and separate from one another? If you wrote down multiple identities, do these identities overlap or intersect? In considering your answers to these questions, also consider *how* and *why*.

Represent

Your identities may be equally valuable and meaningful to you. However, some identities may hold more value and meaning to you than others (Roccas & Brewer, 2002). Reflect on which, if any, of your identities are more important to you. Which are less important to you, if any? Then, create a pictorial representation that depicts the structure (whether hierarchical or equal) of your identities. You can use whatever method you choose to represent your identities, from shapes to pictures, to colors. Be creative! Importantly, understand that this can change over time. There is no "right" or "wrong" answer. Figure 2.2 shows one example of what a pictorial representation might be for my own identities to help guide you.

Reflect

Reflect on your pictorial representation. You might consider sharing your pictorial representation with others—particularly those that assisted you in the "introductory" activity. Is there anything surprising to you about which identities you recognized? Is there anything surprising to you about which identities are most important to you? Is there anything about the pictorial

My Personal
Identities in
Pictorial Form

FIGURE 2.2 A pictorial representation of my own identities. Notice that some identities are displayed as larger than others. These are the identities that I most strongly hold. Those that are displayed smaller are still important, but are not as prominent in my view of who I am within the world at this current moment in time.

Sources: From *Bandage, love, mom icon*, by Muhammad Tajudin, Iconfinder (https://www.iconfinder.com/icons/6169654/bandage_love_mom_mother_day_motherhood_young_mother's_day_icon). CC BY 3.0, from *Pen, write icon*, by khushmeen icons, 2013, Iconfinder (https://www.iconfinder.com/icons/9165599/pen_write_icon). CC BY 4.0, and from *Academic, academy, blacboard* [sic]*, board, chalkboard, education, knowledge, learning, school, student, teach, teaching,...*, by YMB Properties, Iconfinder (www.ymbproperties.com).

representation that you would change so that the depiction is more accurate? If you shared your pictorial representation with others, what do they notice about it? Do you agree or disagree with their perceptions? Why or why not?

Influences

Identity is derived from your observation of others—both of those you want to be like and those that you do not want to be like (Roccas & Brewer, 2002). We reflect on who we want to be (or don't want to be) based on our observations of others as well as our social interactions with others. This allows us to consider which identities we want to adopt and which identities we want to reject. Our adoption of identities is further influenced by the status and privilege that are afforded us by affiliation with particular roles, organizations, and contexts. We tend to adopt those that afford us increased status, social privilege, personal agency, knowledge, and values. Now, reflect on how the interactions that you have had with others, as well as the influences of context, roles, and organizations, have had on your adoption of your identity(ies). How have those interactions influenced your acceptance or rejection of particular identities? How have those interactions influenced how your identities fit within your particular space—especially your degree program and career field?

Tensions and Supports

As I shared earlier in this chapter, identities are malleable and flexible (Miller, 2021). They will (and probably should to some extent) change over time as our experiences change, our goals change, and what we place value on changes. There will undoubtedly be tensions, though, that limit the extent to which our identities are compatible with our life experiences, roles, and responsibilities. Likewise, there will also be supports that foster the extent to which our identities are compatible with our life experiences, roles, and responsibilities. What tensions have you (or do you currently) experienced in relation to your identities? What supports have you (or do you currently) experienced in relation to your identities? How much agency do you have to alter the tensions or supports that you have identified? If you experienced tensions primarily in the past (e.g., when you were a student, when you first entered your career field), what options or abilities did you have (if any) to change anything about those tensions? Note that it's okay if you didn't *do* anything to change those tensions–just focus on what ability or choice you had in the matter.

Reflect

Consider the tensions and supports that you identified as well as the amount of agency that you had (or have) to make changes. Reflect on how those tensions or supports did or didn't (or do or don't) assist you in feeling engaged, empowered, and supported in your role, specifically as a faculty member at your institution. Reflect on how those tensions or supports influence your ability to be an effective guide for students, especially within a mentoring context. What can you do or change in order to be better situated as a role model and guide for students? Draft a plan that includes action steps for better situating yourself within the mentoring context. This plan does not need to be extensive but should include attainable goals that will ultimately

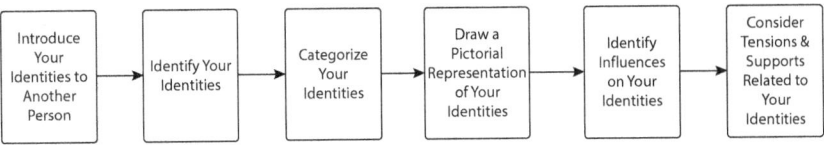

FIGURE 2.3   A graphic representation of the process for faculty identification and development of identities based on a workshop presented by O'Sullivan and Irby (2021).

enhance your ability to support your students while simultaneously attending to your personal well-being (which I'll talk more about in Chapters 4 and 5 as well). See Figure 2.3 for a graphic representation of the entire process that I have outlined for you. Now, follow through on your plan!

Facilitating Identity Development among Students

Now that you have identified and interrogated your own identities, as well as how you might better situate yourself to receive support for honoring your identities, let's talk about how you can support your students' development of identity—particularly within the context of facilitating peer mentoring relationships. While some peer mentoring relationships may be forged organically (such as those involving students connecting with one another informally without faculty intervention), you might have the opportunity to help facilitate a more formal peer mentoring relationship among your students. If the latter is the case, it is helpful for you to understand how students identify and, importantly, to consider identity compatibility as you match students. This does not mean that only students with completely compatible identities can engage in peer mentoring relationships and, thus, should be matched. Finding students with completely compatible identities may not even be possible. Importantly, research also supports the benefit of diversity within the mentoring relationship (Cozza, 2022). It simply means to be cognizant of students' identities, considering the tensions and supports that might be present within the context of the proposed peer mentoring relationship, as you match students. If you are not able to match students for whatever reason, that's okay. Being aware of students' identities and the tensions and supports that influence those identities can still be beneficial.

You can learn about students' identities by having them complete an activity similar to the one above, or you can do so by simply asking students to introduce themselves and by asking follow-up questions that get to the heart of who/what they admire, who/what they aspire to be, who/what they affiliate with, and how they believe that others' perceive them within the world. When asking students about their identities, it's important to be mindful of *how* you go about doing so. You will want to do so in a way that feels safe, welcoming, and non-judgmental. It will be important to acknowledge how your students introduce themselves and the information that they choose to share with you on their identities with openness and unconditional acceptance. The key here is to *listen*. This is not the time to give advice. You are simply listening and gathering information on how your students perceive themselves and how they believe that others perceive them within the world. I also encourage you to read the section on empathic listening in Chapter 4 before really diving into asking questions. You will want to make sure that you truly understand what it means to listen in order for this exercise to be effective!

After learning about students' identities, it will be helpful for you to determine what tensions or supports exist that are impacting the extent to which their identities are compatible with their goals. This also means that you will need to have a conversation of students' goals—what they wish to get out of their current courses and degree program, as well as what they want to get out of the peer mentoring relationship. Be sure to ask students what they hope to gain from your guidance over the peer mentoring relationship as well. Then, work together to determine the level of agency that the student has in making changes to meet their goals by reducing the tensions and increasing the supports that influence their identity compatibility. Chapter 4 will also support you through this process.

After determining the level of agency that the student has, work together to create an action plan. Just as before, this plan need not be extensive, but it will help you identify the types of resources, opportunities, and support that you can provide to the student, as well as others within the peer mentoring relationship. Reflect specifically on what you can do to ensure that the institutional environment is welcoming and honors students' identities. This is especially important when considering barriers that marginalized populations are apt to run into. What can you do to help mitigate those barriers and to create a warm, welcoming, and respectful environment? What, ultimately, do your students need from you in order to both engage in a peer mentoring relationship effectively and meet their academic and career goals?

Summary

In this chapter, we have discussed Identity Theory and why understanding, recognizing, and respecting your own identities—as well as the identities of your students within the peer mentoring relationship--is important.

We have reviewed key points from the research literature that support the use of mentoring in fostering positive identity development. And an exercise on how to interrogate your identities—which can be applied to learning about your students' identities—was shared. As we wrap up this chapter and move forward in our learning, take a few moments to consider the questions and prompts presented in the Reflect and Act section.

Reflect and Act

1 Reflect on the definition of identity provided at the beginning of the chapter. What experiences, beliefs, and other characteristics might influence a person's identity? Why is identity important?
2 Take the time to work through the exercise presented within the chapter on interrogating your own identities.
3 Take some time to fully consider your personal and professional identities. What are they? What has influenced you to develop those identities? How do your identities overlap and intersect? Are there any areas in which your identities are difficult to reconcile? If so, why?
4 Reflect on how your identities might have changed over time. Are your identities now the same or different than they were when you were just beginning your post-secondary education journey? Think about what your identities might have been and why during the life stages that your students are currently experiencing. What perspective does this offer to you?
5 Consider how your students in the peer mentoring relationship might identify. Are there any particular identities that may support their integration into their field of interest? If so, how can you honor and utilize those identities to encourage and support students?
6 Consider how your students in the peer mentoring relationship might identify. Are there any particular identities that may inhibit their integration into their field of interest? If so, how can you honor and support the development of those identities within the particular context of students' field of interest?
7 What steps can you take to more intentionally recognize, respect, and support students' identity development within the context of the peer mentoring relationship, both personally and professionally?
8 What goals have your students set for the peer mentoring relationship? Consider your students' identity compatibility with their goals. Then, consider the level of agency that your students have in making changes to meet their goals by reducing tensions and increasing the supports that influence their identity compatibility. How might you be of support in this process?

References

Atkins, K., Dougan, B. M., Dromgold-Sermens, M. S., Potter, H., Sathy, V., & Panter, A. T. (2020). "Looking at myself in the future": How mentoring shapes scientific identity for STEM students from underrepresented groups. *International Journal of STEM Education, 7*(42). https://doi.org/10.1186/s40594-020-00242-3

Clark, S. L., Dyar, C., Maung, N., & London, B. (2016). Psychosocial pathways to STEM engagement among graduate students in the life sciences. *CBE-Life Sciences Education, 18*, 1–13.

Cozza, B. (2022). Driving theoretical frameworks—women leaders in higher education systems. In B. Cozza & C. Parnther (Eds.), *Voices from women leaders on success in higher education: Pipelines, pathways, and promotion* (pp. 3–14). Routledge.

Daniels, H. A., Grineski, S. E., Collins, T. W., & Frederick, A. H. (2019). Navigating social relationships with mentors and peers: Comfort and belonging among men and women in STEM summer research programs. *CBE-Life Sciences Education, 18*. https://doi.org/10.1187/cbe.18-08-0150

Elmhirst, S. (2024). The rise and fall of the Trad Wife. *The New Yorker*. https://www.newyorker.com/culture/persons-of-interest/the-rise-and-fall-of-the-trad-wife

Erikson, E. H. (1950). *Childhood and society*. Norton.

Garcia-Murillo, Y., Sanchez, B., Carter, J. S., McMahon, S. D., & Schwartz, S. E. (2023). Natural mentoring among college students of color: Considerations for their ethnic-racial identity and psychological well-being. *Journal of Community Psychology, 51*(8), 3348–3365.

Huderson, B., & Huderson, A. (2019). Urban STEM education: A vehicle for broadening participation in STEM. In J. L. Wendt & D. L. Apugo (Eds.), *K-12 STEM education in urban learning environments* (pp. 1–24). IGI Global.

Hurtado, S., Chang, J. C., Saenz, V. B., Espinosa, L. L., Cabrera, N. L., & Cerna, O. S. (2007). Predicting transition and adjustment to college: Minority biomedical and behavioral science students' first year of college. *Research in Higher Education, 48*(7), 841–887.

Ivey, S. S., & Parish, S. G. (2021). Cultivating STEM identity through the peer mentoring relationship. In A. J. Rockinson-Szapkiw, J. L. Wendt, & K. Wade-Jaimes (Eds.), *Navigating the peer mentoring relationship: A handbook for women and other underrepresented populations in STEM* (pp. 153–165). Kendall Hunt Publishing Company.

Jones, V. O., & Wendt, J. L. (2025). Encouraging confidence: The impact of an online peer mentoring program on women peer mentees in STEM at two HBCUs. *Trends in Higher Education, 4*(3). https://doi.org/10.3390/higheredu4010003

Love, B. L. (2019). *We want to do more than survive: Abolitionist teaching and the pursuit of educational freedom*. Beacon Press.

Miller, D. (2021). *Honoring identities*. Rowman and Littlefield.

Moore, E. Jr., Michael, A., & Penick-Parks, M. W. (2018). *The guide for White women who teach Black boys*. Corwin.

Murrell, A. J., Blake-Beard, S., & Porter, J. D. M. (2021). The importance of peer mentoring, identity work and holding environments: A study of African American leadership development. *International Journal of Environmental Research and Public Health, 18*(9). https://doi.org/10.3390/ijerph18094920

National Science Foundation. (2023). *Diversity and STEM: Women, minorities, and persons with disabilities*. National Center for Science and Engineering Statistics. https://www.nsf.gov/reports/statistics/diversity-stem-women-minorities-persons-disabilities-2023

National Science Foundation. (2024). *The STEM labor force: Scientists, engineers, and skilled technical workers.* https://ncses.nsf.gov/pubs/nsb20245

O'Sullivan, P. S., & Irby, D. M. (2021). Educator identity formation: A faculty development workshop. *MedEdPORTAL, 17.* https://doi.org/10.15766/mep_2374-8265.11070

Roccas, S., & Brewer, M. (2002). Social identity complexity. *Personality and Social Psychology Review, 6*(2), 88–106. https://doi.org/10.1207/S15327957PSPR0602_01

Rockinson-Szapkiw, A., Sharpe, K., & Wendt, J. L. (2022). Promoting self-efficacy, mentoring competencies, and persistence in STEM: A case study evaluating racial and ethnic minority women's learning experiences in a virtual STEM peer mentor training. *Journal of Science Education and Technology, 31,* 386–402. https://doi.org/10.1007/s10956-022-09962-3

Rockinson-Szapkiw, A., & Wendt, J. L. (2020). The benefits and challenges of a blended peer mentoring program for women peer mentors in STEM. *International Journal on Mentoring and Coaching in Education, 10*(1), 1–16.

Saffie-Robertson, M. C. (2020). It's not you, it's me: An exploration of mentoring experiences for women in STEM. *Sex Roles, 83,* 566–579.

Society of Women Engineers. (2023). *Earnings gap.* https://swe.org/research/2023/earning-gap/

Taylor, K.-Y. (2017). *How we get free: Black feminism and the Combahee River Collective.* Haymarket Books.

Wendt, J. L., & Jones, V. O. (2024a). Peer mentors' experiences in an online STEM peer mentoring program: "Beacons of light." *International Journal of Mentoring and Coaching in Education, 13*(3). https://doi.org/10.1108/IJMCE-03-2023-0033

Wendt, J. L., & Jones, V. O. (2024b). Supporting BIPOC males in STEM: Insights from a case study on online peer mentoring. *Journal of Research in STEM Education, 10*(1–2), 89–113. https://doi.org/10.51355/j-stem.2024.145

Wilton, M., Katz, D., Claremont, A., Gonzalez-Nino, E., Foltz, K. R., & Christoffersen, R. E. (2021). Improving academic performance and retention of first-year biology students through a scalable peer mentorship program. *CBE-Life Sciences Education, 20,* 1–13.

Zajac, T., Magda, I., Bozykowski, M., Chlon-Dominczak, A., & Jasinski, M. (2025). Gender pay gaps across STEM fields of study. *Studies in Higher Education, 50*(1), 126–139.

3
CULTURAL RESPONSIVENESS, MOTIVATION, AND GROWTH MINDSET

Introduction

In Chapter 2, I shared with you the importance of understanding and honoring your identities as well as your students' identities, especially within the context of supporting peer mentoring relationships. As you work to honor identities with integrity, you will inevitably find yourself needing to evaluate your personal values while also respecting the values that your students hold. Take a moment to consider what you personally place value on. What is most important to you in the grand scheme of things? What is most important to you professionally? For instance, do you place more value on gaining greater prestige within your field of expertise, or is there a different, potentially more meaningful legacy that you wish to leave behind? Do you place more value on publishing prolifically and disseminating novel findings before someone else "beats you to it," or do you place more value on affording students the opportunity to collaborate and learn about the publishing process? Do you place more value on simply holding office hours, or do you place more value on cultivating meaningful relationships with students that foster the skills necessary for independent problem-solving and lifelong learning? (Note that these questions are not meant to be presented with judgment. Rather, they are presented as one way to interrogate what is ultimately most important to you as related to your role within the realm of education.)

Take some time now to reflect on your values, the activities that give meaning to not only your career but your life as a whole, and what you wish to achieve in your current faculty role in alignment with those values and activities. Write these thoughts down. Then, we'll explore what it means to act in alignment with your values, which, inherently, will inform your ability to guide students in a manner that is culturally responsive.

Cultural Responsiveness

Cultural responsiveness seems to be pervasive within educational reform efforts and the literature across the field of education today. Especially within the current political climate, it seems that cultural responsiveness is at the forefront of most conversations related to the education landscape—the news, popular media. But what does it actually mean to be culturally responsive? Why is it important? How do faculty members make sure that they are cultivating an environment that is culturally responsive for their students? What does that look like in the context of supporting peer mentoring relationships? Let's explore this further.

Understanding cultural responsiveness begins by (a) recognizing the historical context upon which our education system was founded, thus influencing societal context and experiences and (b) recognizing and embracing the value in students' uniquely individual histories, experiences, identities, and cultures. If we critically and honestly reflect on the educational landscape historically—especially within the United States (US)—it is evident that the education system as a whole was structured and implemented in a way that favored the majority population, which has historically been White male students (Malott, 2021; Miller & Walker, 2023; Wells, 2024). It is a simple fact, although an increasingly contentious point, especially in the current political climate, but one that is supported by history (Miller & Walker, 2023). And, perhaps just as importantly, this structure persists today (Love, 2019; Malott, 2021; Miller & Walker, 2023; Milner, 2021; Morris, 2018; Wells, 2024).

Malott (2021) (as well as many others) provides an in-depth explanation of the historical context of education within the US, which I encourage you to explore (see Chapter 7 for additional resources). For the sake of brevity, though, I'll give you a quick summary beginning with colonial times just to set the stage. (Again, I encourage you to explore the nuances through historically-focused texts.) During colonial times, ministers were first charged with providing education to the public, which consisted mostly of religious services and scripture (Bible) reading (Malott, 2021). These teachings were based on Roman Catholic beliefs and reinforced the idea of devotion to the colony and fear of the supposed barbarism of the Indigenous peoples whose land was stolen so that the colonies could be built in the first place. Meanwhile, education was also used to convert Indigenous folks and others who were forcibly brought to the colonies (the enslaved) to Christianity—but not to advance their knowledge in key skills such as reading and writing (which they were forbidden to learn).

Eventually, the focus of education shifted to a more capitalistic approach with the understanding that an educated populace would result in a stronger workforce (Malott, 2021). Grammar schools, the first public schools in the United States, were created but could only be attended by White male students who had sufficient financial resources (Powell, 2012), thus positioning While men well for jobs and accumulation of wealth. Women were generally not allowed to attend, and any student who identified as any race or ethnicity other than White was similarly excluded (Love, 2019; Miller & Walker, 2023). In rare exceptions, women were allowed to attend dame schools, but it wasn't until the mid-1800s that women were more broadly provided with the opportunity to engage in formal education (Chamberlain, 1988; Gale, 2025).

Similarly, in the late 1700s, the first public schools for Black students were established (Kates, 2017); thus, finally affording Black folks the opportunity to engage in formal education. It wasn't until 1954 with *Brown v. Board of Education* that segregation in public schools was legally ended (Supreme Court of the United States, 1954), although the extent to which segregation actually came to an end is debatable given the disparities in funding, resources, and support based on geographic location and, largely, student population demographics that persists in the US today (Love, 2019; Morris, 2018; Wells, 2024).

In the 1800s, the first off-reservation Indian boarding schools were established, which led to the forcible removal of Indigenous children from their homes and families for the purpose of committing cultural genocide (Coalition, 2025; Mejia, 2025). The primary purpose of these schools was to force Indigenous children to assimilate to the culture and customs of the White majority (Mejia, 2025). Thus, the use of Indigenous language and practicing of Indigenous culture was expressly forbidden and punishable, ending in not only loss of culture but also pervasive abuse and even murder (Coalition, 2025; Kimmerer, 2013; Mejia, 2025). While the practice of forcible removal of Indigenous children to attend boarding schools came to an end in the mid-1900s, a few Indian boarding schools still exist today—although they are now within the control of the Indigenous community (Mejia, 2025).

This historical context, although very briefly summarized, is partly where a deficit-minded view within education was instilled (Wells, 2024), where students who were perceived as "different"—typically anyone other than those within the majority White male population—have been perceived as lacking in some way. Latinx students, Black students, and Indigenous students have all been perceived at one time or another within history as being "less than." Often, the perception was that these students were not as intelligent as their White counterparts, or didn't receive appropriate attention, instruction, and resources at home. Likewise, the perception was that these individuals experienced lower socioeconomic status and, thus, were bound to have limited education experiences. Still somewhat pervasive, the perception has also historically been gender-biased—that girls couldn't be capable of engaging competently in mathematics or science. Any number of perceptions could be inserted into these statements, representing historical thought that has been proven time and time again to be inaccurate, misguided, and downright hurtful to students socially, emotionally, and academically. Each of these assumptions is preposterous, right?

Unfortunately, the deficit-minded view is still quite pervasive (Wells, 2024). Importantly, even educators—those who are trained and trusted to be the experts in education—are not always presented with an accurate reflection of the historical context of education within the US (Miller & Walker, 2023; Wells, 2024), which leads to an inaccurate understanding of the structures upon which the education system has been built. This, in turn, often leads to educators perpetuating inequity within their own classrooms, even unintentionally. While I could devote an entire library of books to this topic, what I want to focus on within the context of peer mentoring is this: The current push for cultural responsiveness is centered on recognition of the immense value and richness that is interwoven within each student's cultural experiences, values, desires, and thus, their lives. All students bring value—this is not limited to one particular race, ethnicity, culture, socioeconomic level, gender, or sex. And all students have exponential possibility. It is your role as a faculty facilitator to intentionally interrogate your own unconscious biases, set those biases aside, and adopt a worldview that sees value, talent, and ability in *all* of your students. But it's difficult to adopt this view if you aren't able to see what has gone wrong historically and what factors are still at play within today's academic arena.

For you as the faculty facilitator, this means recognizing that students' ability to achieve academically goes far beyond the curriculum taught. It is your beliefs as the faculty facilitator that will ultimately drive culturally responsive pedagogy (Ladson-Billings, 2021), especially within the context of assisting your students within the peer mentoring relationship with discipline area content (academic-focused mentoring). Ladson-Billings, a renowned leader in culturally responsive teaching practices, shares that it is educators' "beliefs about themselves and others; about how to structure social relations; and about the nature of knowledge" (2021, p. 3) that form the foundation of cultural responsiveness (see Figure 3.1). This extends to you as the faculty facilitator as well.

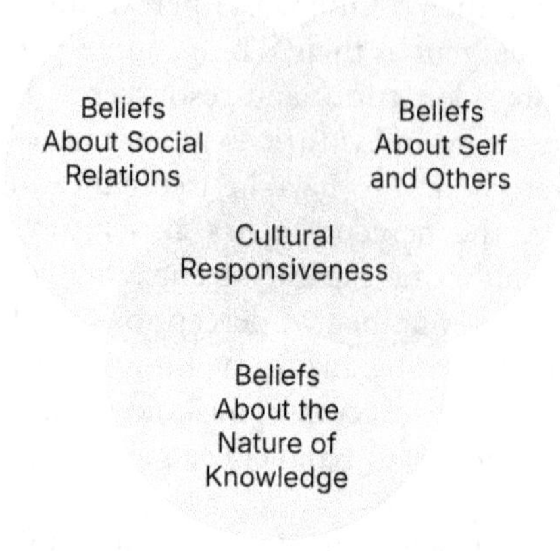

```
         Beliefs                    Beliefs
      About Social                 About Self
        Relations                  and Others

                    Cultural
                 Responsiveness


                     Beliefs
                   About the
                   Nature of
                   Knowledge
```

FIGURE 3.1 The foundation of cultural responsiveness as described by Ladson-Billings (2021).

It is these beliefs that allow you as the faculty facilitator to move from thinking solely about discipline area knowledge to thinking about how to facilitate opportunities for equitable student learning through your role in the peer mentoring relationship (Zhou, 2021). It is these beliefs that help you as the faculty facilitator consider how to cultivate your own cultural competence, allowing for a more robust understanding of experiences that are very likely different from your own. It is these beliefs that will drive you to support the development of both sociopolitical consciousness and critical consciousness (Ladson-Billings, 2021). "Ultimately the context of mentoring *is* difference: Who we are shapes our thinking, our conversation, our relationship, and our behaviors" (Zachary & Fain, 2022, p. 37). With this in mind, the three primary components of culturally relevant pedagogy are academic achievement, cultural competence, and sociopolitical or critical consciousness (See Figure 3.2) (Ladson-Billings, 2021).

When interrogating my own abilities to engage in cultural responsiveness, I personally have found it helpful to engage in open and honest conversations with a trusted colleague that shares a different background than my own. This has allowed me to "see" my actions as an educator through a lens other than my own and to more accurately assess where I might hold biases that otherwise would remain hidden. While it is not any other person's responsibility to teach me (or you!) how to be culturally responsive to another group, each of us can engage in self-reflection and ask for constructive feedback to shed light on areas of self that we may not easily be able to recognize.

I have also found it helpful to intentionally engage in events that highlight cultures and experiences that are different from my own, engaging as

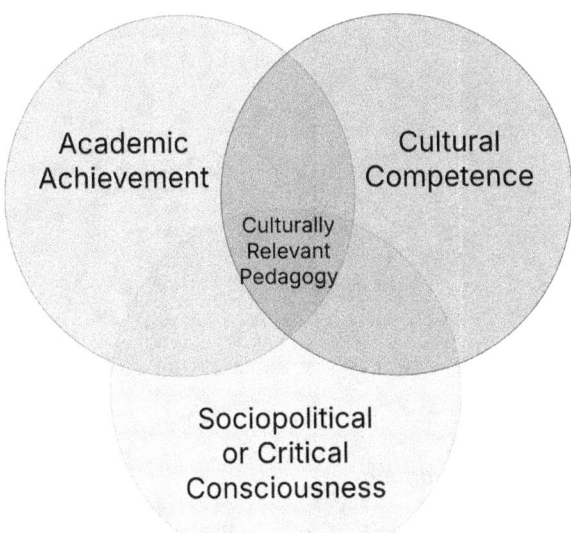

FIGURE 3.2 The three primary components of culturally relevant pedagogy as described by Ladson-Billings (2021).

a curious bystander, a vessel for obtaining knowledge that otherwise would go unfilled. Likewise, as an avid reader, I have found great value in reading books—fiction and non-fiction alike—written by authors with different cultural, racial, and ethnic backgrounds and gender identities than my own. The idea is not to judge or perceive one way as right or one way as wrong, one way as better or another way as worse, but to *learn*. Respect can only bloom through exposure and receptiveness. The experiences of others can bring insight into the world around you that you otherwise would not be privileged to experience.

As a faculty facilitator, it will be necessary for you to engage in critical thought surrounding your own values, perspectives, and experiences. Take a moment to pause and consider. What are your values? What is most important to you? What are your core beliefs about self? Others? What are your core beliefs about your role as a faculty member? What about students' roles? Cultural responsiveness revolves around the idea that students' academic success should not come "at the expense of their cultural and psychosocial well-being" (Ladson-Billings, 2021, p. 26). Thus, in order to be a culturally responsive faculty member, it is essential that your actions ultimately must align with and be consistent with your own beliefs and values (Ladson-Billings, 2021). This is especially important within the context of supporting students engaging in peer mentoring relationships (Figure 3.3).

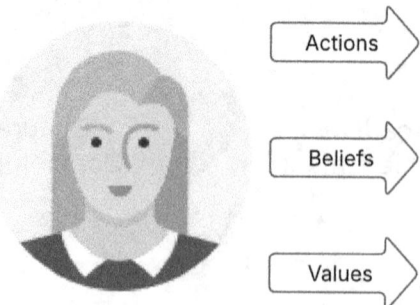

FIGURE 3.3 In order to be culturally responsive, your actions, beliefs, and values must align.

Culturally Responsive Pedagogy

As a faculty facilitator, you are also an educator—a teacher. Thus, having an understanding of culturally responsive pedagogy is integral to actually engaging in culturally responsive practices. Culturally relevant pedagogy can be defined as a practice of teaching "that empowers students intellectually, socially, emotionally, and politically by using cultural referents to impart knowledge, skills, and attitudes" (Ladson-Billings, 2021, p. 4). What this looks like at any given time within any given classroom or laboratory with any given group of students might differ—and quite frankly *should* differ. Each of your students is unique, as are their experiences, their values, their knowledge, and their needs. Thus, as a culturally responsive educator practicing culturally relevant pedagogy, you will need to take the time to first get to know your students—especially those whom you are guiding through a peer mentoring relationship. Who are they as individuals? What experiences do they bring to the table? How might these experiences shape the way in which they view and participate in the world around them? How might these experiences shape their attitudes? What knowledge do they have that can both be built upon and shared with those around them? What about the world is important and of value to your students? Why?

You can get to know your students by first establishing a safe and welcoming environment for them. Within the context of supporting peer mentoring relationships, this obviously extends beyond the bounds of the classroom to the laboratory, your office, institutional spaces in which you'll interact, and beyond. Consider the ways in which you, your presence, and your approachability might be perceived by students. Are you perceived by students as approachable? Why or why not? Do you interact with students in a non-judgmental manner, accepting who they are as individual human beings? Is this conveyed to students and, if so, how? If not, why? In what ways are students able to interact and engage with you? For instance, do you offer multiple avenues for connection that might fit different levels of need, comfort, and accessibility (e.g., email, face-to-face office hours, phone)? What tone do you set when interacting with students individually and within peer mentoring groups? Is that tone conducive to inviting students to share their authentic selves? Are there any potential stereotypes or biases that you may be perpetuating within these spaces, even unintentionally?

Pay attention to what students say to you in your daily conversations. How do they say it? What does their body language (non-verbal cues) tell you about their level of comfort in interacting with you? Oftentimes, our initial reaction to student communication focuses on whether the communication is "respectful" or "appropriate." Instead of jumping to judgment of perceived intent and reacting immediately, take a moment to consider the *why* behind the *how*. That is, why might a student communicate with you (or their peers) in any particular way? Their verbal language as well as non-verbal language (body language) often tells a more complete story about their experiences and current feelings of acceptance within a particular space. Consider how their past experiences and individual culture might shape how and what they communicate. Are there ways that you can facilitate more open and welcoming communication? (We'll explore methods for effective communication a bit more in Chapter 6.)

Being *reflective* rather than *reactive* is essential. It is imperative that you as a faculty facilitator engage in regular, intentional self-reflection. This can take many forms from journaling to conversing with a trusted colleague to simply thinking about your actions and words during a quiet time at the end of the day. Ultimately, we are all human and our experiences and contexts by which we are surrounded will influence our thoughts, words, and actions. But through deliberate work, we can all work toward a more inclusive, equitable education environment—which inevitably includes the peer mentoring environment. Use the questions in the Reflect and Act section at the end of this chapter to help guide you.

So, now that you have a more robust understanding of cultural responsiveness, let's talk about motivation.

Motivation and the "Push" to Succeed

As a faculty member, it is highly likely that you have completed many courses, earned degrees, and perhaps even earned additional certifications. I likely do not have to remind you that inherent to those accomplishments is some degree of motivation. The concept of motivation, then, should be relatively familiar! Understanding what motivates our students, though, may take some creative thinking, observation, and a lot of listening.

Carol Dweck shares that "motivation is at the core of human psychology" (2017, p. 689). Motivation encompasses our drive, our internal "push" to do the hard things and move toward accomplishing some goal. It is what "'moves' people to action" (Ryan & Deci, 2017, p. 13). It is "what *energizes* and *gives* direction to behavior" (Ryan & Deci, 2017, p. 13). Motivation is influenced by myriad factors—both internal and external. For instance, students might be motivated based on their own core values–thus, internal motivation. Or students might be motivated because accomplishing a task will yield an external reward, such as higher pay.

But how exactly does motivation work? What drives a student to become motivated? And how can faculty use those factors to support students' motivation and, likewise, persistence when interacting through the peer mentoring context? I don't want to get too far into the weeds, but in order to answer these questions, it is helpful to understand some of the underlying theory that frames motivation. Please understand, however, that the following descriptions are merely a snapshot of the most common contemporary theories that explain the concept of motivation. There are myriad theories that address motivation. I urge you to explore more if you find this topic of interest to you. Some additional resources are listed in Chapter 7.

A few of the more current contemporary theories related to motivation include the Attribution Theory of Motivation (Weiner, 1985), Situated Expectancy-Value Theory (Eccles & Wigfield, 2020; Wigfield & Eccles, 2000), Self-Determination Theory (Ryan & Deci, 2017), and Social-Cognitive Theory (Bandura, 1986). These theories in particular have been closely tied to motivation in science and engineering within the research literature (The National Academies of Science, Engineering, and Medicine, 2019). Let's take a brief look at each. (For a more robust description of each theory, see the additional resources in Chapter 7.)

Attribution Theory of Motivation

Using Atkinson's (1957) particular views of Expectancy-Value Theory as a foundation, Weiner (1985) proposed the Attribution Theory of Motivation (as a sidenote, Atkinson was Weiner's mentor). The Attribution Theory of Motivation suggests that, when individuals experience the same event or circumstance yet respond in different ways, their differing responses are a result of their unique perceptions of the cause of the outcome. In this theory, there are three dimensions that can determine or influence a cause: (a) the locus/root of the cause (e.g., whether the cause is internal or external to the individual), (b) the degree of stability of the cause (e.g., the degree to which the cause is stable or malleable), and (c) the degree of control of the cause (e.g., the degree to which the individual can exert control

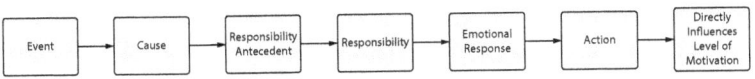

FIGURE 3.4 A very simplified visualization of the Attribution Theory of Motivation.

over the cause) (The National Academies of Science, 2019; Weiner, 1985; Weiner et al., 2012).

This theory is helpful for understanding why individuals in similar circumstances experience different outcomes and, importantly, generate different emotions, which can in turn influence their levels of persistence and motivation. Weiner and colleagues (2012) explain that when an individual believes that their success at a given task can be attributed to self, then they experience an enhanced feeling of pride. On the other hand, if they believe their failure at a given task can be attributed to a controllable aspect of self (such as a lack of effort, for instance), then they experience an enhanced feeling of guilt. If their failure at a given task can be attributed to an uncontrollable aspect of self (such as lack of intelligence), then they instead experience an enhanced feeling of embarrassment, shame, or humiliation. If an individual attributes the outcome of an experience to luck, they experience a feeling of surprise, no matter whether the outcome was positive or negative. If the outcome of an experience is attributed to an external cause (such as to other people), the individual experiences feelings of anger. These causal effects continue depending on exactly how the individual perceives the cause of an outcome to exist and can be further visualized in a very simplified diagram (see Figure 3.4). And, of course, how an individual feels directly influences whether they are motivated to continue at a given task or whether their success is viewed as hopeless.

Situated Expectancy-Value Theory

Situated Expectancy-Value Theory (Eccles & Wigfield, 2020; Wigfield & Eccles, 2000) posits that the choices that an individual makes to engage in a particular task (which could be an individual activity or a particular career choice), their level of engagement in the task, and degree of success at meeting a desired outcome are all determined by the value that they place on the task in combination with their expectancy of experiencing success. The value that an individual places on a task is related to their enjoyment of the task (also called intrinsic value), which directly relates to their level of motivation to continue engaging in the task. When an individual places a high intrinsic value on a task, they will be more motivated to engage with the task and more likely to persist (see Figure 3.5).

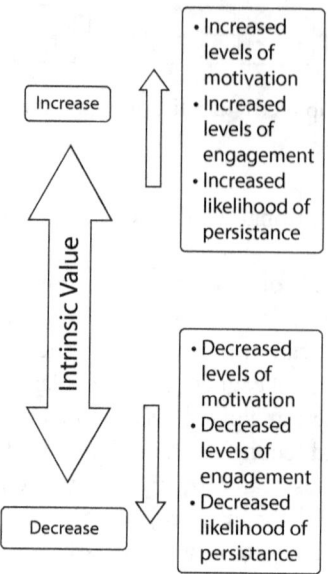

FIGURE 3.5 The relationship between intrinsic value and motivation, engagement, and persistence according to the Situated Expectancy-Value Theory.

Self-Determination Theory

Self-Determination Theory is one of my favorite theories because it really gets to the heart of the many factors that influence an individual's behavior and actions. Self-Determination Theory is primarily focused on the level of success an individual can achieve or, in other words, how well they can flourish or thrive (Ryan & Deci, 2017). The theory attempts to explain what individuals need from their surrounding environments (both psychologically and socially) in order to thrive. Behavior is seen as a function of motive with motive being influenced by the values that an individual holds. The premise of the theory is that there are three basic psychological needs: autonomy, competence, and relatedness. *Autonomy* can be viewed as an individual's ability to self-regulate their experiences and actions, which typically align with their values and interests. *Competence* can be thought of as level of mastery, and *relatedness* can be thought of as social connectedness. *Internal and external forces* constantly act on each of these three needs, thus influencing one's actions and behaviors. The degree to which the basic psychological needs are met, and the sources through which this occurs (for instance, whether through one's own volition through matching of one's interests and values or whether through coercion by some outside source), influence an individual's

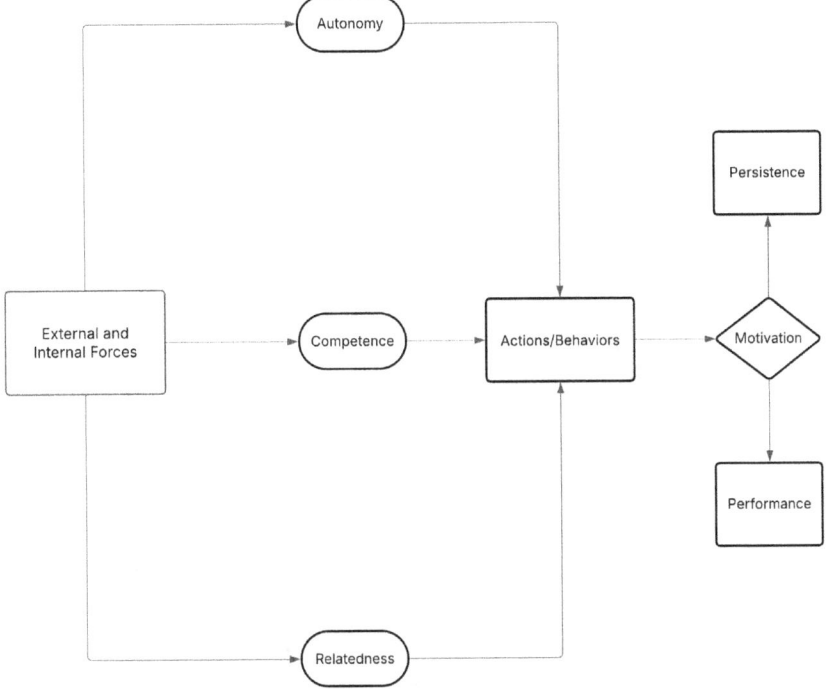

FIGURE 3.6 A simplified visualization of the Self-Determination Theory where external and internal forces influence the three basic psychological needs, which in turn influence motivation, persistence, and performance.

level of motivation. Thus, persistence and performance are ultimately influenced by the types of and sources of motivation (see Figure 3.6).

Social-Cognitive Theory

Social-Cognitive Theory was proposed by Bandura (1986). In this theory, the underlying premise is that an individual has an active role in their experiences. Through observation, an individual can engage in vicarious learning and, thus, reproduce or mimic particular behaviors that lead to a specific outcome—but only when they are engaged, attentive, and self-motivated. Self-motivation leads to self-regulation in that an individual will then engage in purposeful action. They then anticipate a particular outcome, conceptualizing outcomes of specific behaviors, and then set goals to reach their desired outcome. The individual then forms a judgment about their ability to engage in action to experience their desired outcome, otherwise known

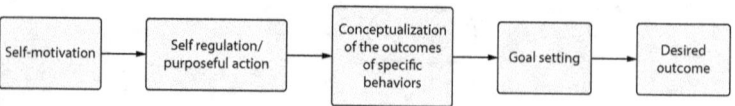

FIGURE 3.7 A representation of Social-Cognitive Theory.

as self-efficacy (Bandura, 1977). Self-efficacy, in turn, impacts an individual's thoughts and actions, including the amount of effort they are willing to put forth. Thus, self-efficacy directly impacts motivation (see Figure 3.7). Further, self-efficacy has been shown to correlate with individuals' choice to pursue STEM activities; in particular, persistence, and motivation (Zeldin & Pajares, 2000; Zeldin et al., 2008). The higher the level of self-efficacy, the more an individual is motivated to engage with and persist in STEM-related activities (e.g., tasks, degree programs, careers).

Tinto's Model of Student Motivation and Persistence

I've found Tinto's (2017) Model of Student Motivation and Persistence to be especially helpful in understanding several of the factors that influence motivation as it ties many of the constructs that are included in the other theories that I've discussed together very nicely. Tinto's model suggests that various factors influence motivation, which itself is malleable and will ebb and flow depending on the experiences an individual is subjected to. Specifically referring to students enrolled in a college or university, students' motivation can be defined as the "outcome of the interaction among student goals, self-efficacy, sense of belonging, and perceived worth or relevance of the curriculum" (p. 255). Thus, it is the interaction of these factors that ultimately influences whether a student will be motivated enough to persist in their studies and, ultimately, meet their goal of degree completion. An increase in a student's sense of belonging and self-efficacy, for instance, will lead to an increase in motivation, which in turn will lead to an increase in their intent to persist. On the other hand, a decrease in their sense of belonging, for instance, may lead to a decrease in their motivation and, thus, a decrease in the likelihood that they will persist (Figure 3.8).

Growth Mindset

Along with understanding contemporary theories that help explain the factors that influence motivation, it is also important to understand concepts related to mindset. Each of us has the "capacity for lifelong learning and brain development" (Dweck, 2006, p. 20). That is, none of us is born with a specific

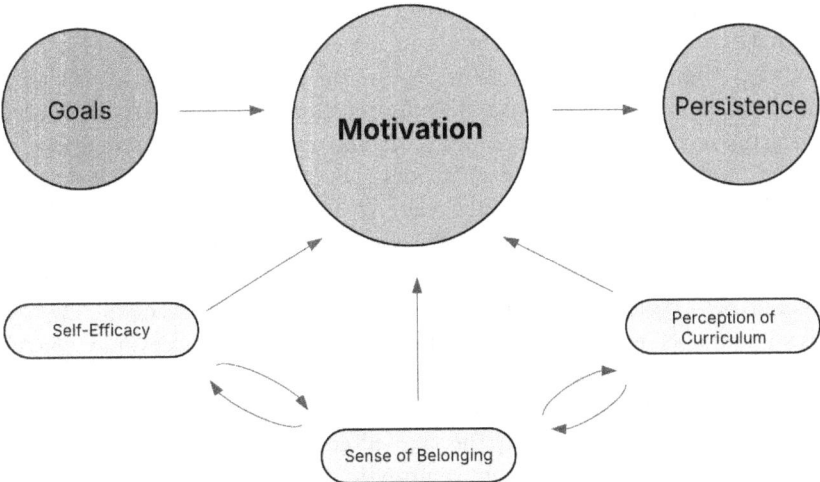

FIGURE 3.8 A visualization of Tinto's Model of Student Motivation and Persistence.

Source: Adapted from Tinto (2017, p. 256).

capacity or limit to what we can learn. Thus, *"the view you adopt for yourself profoundly affects the way you lead your life"* (Dweck, 2006, p. 21). This view determines whether you will become who you want to be, do the things that you want to do, and accomplish the goals that you wish to accomplish.

There are two types of views or mindsets: fixed mindset and growth mindset (Dweck & Yeager, 2019; Elliott-Moskwa, 2022). A fixed mindset is present when an individual believes that they have little to no control over a particular characteristic or ability. Thus, their ability is viewed as static and "carved in stone" (Dweck, 2006, p. 21). A growth mindset, on the other hand, is present when an individual believes that they begin with a particular level of a characteristic or ability but can learn and grow over time. Their ability is seen as malleable and "is based on the belief that [their] basic qualities are things [they] can cultivate through [their] efforts, [their] strategies, and help from others" (Dweck, 2006, p. 23). Individuals with a growth mindset are more likely to be resilient in the face of challenges and to persist in meeting their goals—they are learners (Dweck, 2006). Those with a fixed mindset, however, tend to avoid challenges whenever possible and are, thus, less likely to persist in meeting their goals—especially when meeting those goals is difficult or outside of their comfort zone.

There is a substantial amount of research that demonstrates the benefits and potential of adopting (or fostering among others) a growth mindset (Dweck & Yeager, 2019; Elliott-Moskwa, 2022), which I encourage you to explore (see Chapter 7 for additional resources). I'll share some of the key points here, though, as it is helpful to have a basic understanding of growth mindset prior to engaging with students in the peer mentoring relationship.

When students adopt a growth mindset, they are more likely to experience positive academic outcomes (e.g., higher grades), resilience, persistence, and challenge-seeking behaviors (Dweck, 2006; Dweck & Yeager, 2019). These outcomes directly relate to their level of motivation—particularly intrinsic motivation, which is encouraged through adopting the belief that one can continuously grow and learn. Importantly, the adoption of a growth mindset also influences the development of academic identity as students become more confident in their abilities to successfully engage with and belong in a particular disciplinary space (Moore et al., 2018) (refer back to Chapter 2 for our discussion on identity development).

Study has found that, in the higher education STEM classroom in particular, professors can support growth mindset by following process-focused teaching practices (such as adapting instruction to student needs and consistently monitoring student progress), intentionally communicating positivity surrounding the amount of effort students expend (even if they still make mistakes or fall short of mastery), and providing process praise (such as communicating positivity around students' use of problem-solving skills, thought processes, progress, and development) (Dweck & Yeager, 2019). These actions can easily be adapted and applied to the peer mentoring context, where you as the faculty facilitator can support students' development of growth mindset academically, socially, and psychosocially with similar benefits.

Dweck shares that "people with the growth mindset thrive when they are stretching themselves" (2006, p. 51). Individuals with a growth mindset seek out new experiences, confront their shortcomings, use mistakes and failure as valuable learning experiences, apply feedback (both from experiences and from others), and alter their strategies to ensure forward momentum toward meeting their goals (Dweck, 2006). What's important, though, is that growth mindset can be taught, and it can be learned! Our mindset—whether growth or fixed—is a very important part of who we are, but it is something that can be changed over time. It's also important to recognize that a person can have a growth mindset in one area and a fixed mindset in another and that this, too, can change based on experiences, context, and intent (Dweck, 2006).

Thus, you as the faculty facilitator can assist students by creating an environment that teaches growth mindset. Keep in mind that "the growth mindset is the belief that abilities can be cultivated" (Dweck, 2006, p. 101). So, as a faculty facilitator, one major way that you can facilitate the adoption of a growth mindset among your students is by focusing on the effort that your students are putting forth. Effort isn't everything, but those that continue to put forth effort even in the face of challenges can experience "transformative power" (Dweck, 2006, p. 86). Effort allows our students to *experience* failure without *becoming* a failure. Effort allows our students to learn from mistakes and grow. The effort that is inherent to a growth mindset "allows people to value what they're doing *regardless of the outcome*" (Dweck, 2006, p. 105). Thus, rather than praising ability, praise your students' efforts–especially when they persist in doing what it takes to succeed. Be intentional about sending messages about process and growth, keeping in mind that every word you utter and every action that you take sends a message to your students (Dweck, 2006). Encourage the development of your students' potential. Encourage your students to seek out successful others as models (you can also model the adoption of growth mindset!). Use setbacks as learning opportunities. Assist your students in taking charge of their forward momentum. By doing so, you will be setting your students up for adoption of a growth mindset and, thus, future success.

Connecting It All

I know that I have presented a lot of information, so let's connect all of the dots within the specific context of facilitating peer mentoring relationships. How do you as a faculty member support students as they engage in the peer mentoring process while simultaneously considering (a) cultural responsiveness, (b) motivation, and (c) growth mindset? As you cultivate a culturally responsive space as well as culturally responsive practices, your students will become more comfortable in interacting authentically with you and within the peer mentoring relationship (see Figure 3.9). When

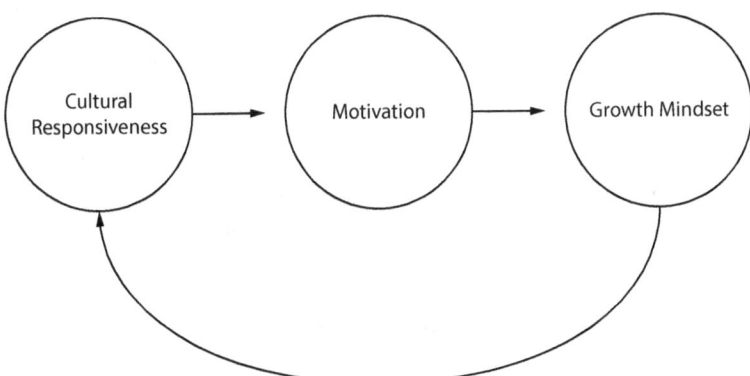

FIGURE 3.9 A visualization of how cultural responsiveness influences motivation, which in turn influences cultivation of growth mindset.

they are comfortable sharing their authentic selves, they will be able to focus more of their time and energy on working toward their goals rather than devoting precious resources to trying to fit in. This will, in turn, encourage motivation. As comfort and motivation increase, their beliefs in their ability to learn and grow will inevitably be enhanced. And, in turn, when they are able to adopt a growth mindset, it is likely that they will then want to support others, which leads them to be more aware of developing welcoming spaces and so on. The cycle continues with each iteration producing support for others and reciprocity.

Keep in mind the value of interrogating and recognizing your own core beliefs and values as you facilitate this process. Your core beliefs and values will directly influence your actions, which will determine how you provide support to students as they navigate the institutional environment—whether academically, socially, or psychosocially. Values alignment in particular is a key component of the mentoring relationship (Ferris & Waldron, 2022). The support and guidance that you provide to students must demonstrate cultural integrity. That is, your actions must align with your beliefs and values, and those beliefs and values must inherently demonstrate respect for others whose experiences, cultures, beliefs, and values may be different from our own. For most, this is a practice that must be continuously attended to, fine-tuned, and adjusted as we each individually learn and grow from new experiences and interactions with those around us. So, give yourself some grace as you work on this. Remember that we all have much to learn and much to give.

Summary

In this chapter, we discussed the importance of interrogating our own values, core beliefs, and assumptions about the world around us, including the people with whom we interact. A very briefly summarized history of education within the United States was presented in an effort to explain the deficit-focused mindset that is common within our educational systems. We explored the concepts of cultural responsiveness and culturally responsive pedagogy, and myriad questions were posed for you to use as a guide in self-reflection. Motivation and the various theories that serve as a foundation for motivation were explained, including Attribution Theory of Motivation, Situated Expectancy-Value Theory, Self-Determination Theory, and Social Cognitive Theory as well as Tinto's Model of Student Motivation and Persistence. We also examined the different types of mindsets, including a fixed mindset and a growth mindset, and the importance of cultivating a growth mindset was shared. Finally, you were encouraged to engage in continuous self-analysis and self-reflection to ensure your own personal and professional growth in creating welcoming, safe environments for your students as you facilitate the peer mentoring relationship. As we wrap up this chapter and move forward in our learning, take a few moments to consider the questions and prompts presented in the Reflect and Act section.

Reflect and Act

1. What are your core beliefs about self and others—especially students who might have different backgrounds or experiences than your own?
2. Have you adopted the belief that all students are capable of success? If yes, wonderful! If not, why not?
3. What values are most important to you?
4. Do your beliefs and values align with your actions? If so, how? If not, how might you be more intentional about doing so?
5. Reflect on your cultural behaviors and cultural assumptions. Take some time to write down your cultural assumptions and reflect on how specifically they are reflected in your actions and behaviors. For instance, a cultural assumption might be that students, like children, should be seen but not heard. This might be reflected in the behavior of talking over students and not using empathic listening when they express their questions or concerns. Do your cultural behaviors and cultural assumptions align with the type of environment and experience you want to provide for your students?
6. Consider the norms of your culture as compared to the norms of your students' cultures. What are the similarities? What are the differences? How might you be mindful of difference in norms?
7. Complete a cross-cultural mentoring inventory. There are several free examples available online through a simple search. Suggestions are provided in Chapter 7.
8. How do you support students' navigation of the institutional environment as it relates to cultural similarities and differences?
9. What theories of motivation most resonate with you? Why?
10. What challenges to motivation have you experienced? How can you use these experiences to help support students' levels of motivation?
11. Reflect on ways that you can cultivate motivation among your students, specifically within the peer mentoring relationship. Are there particular ways that you can model motivation?
12. Reflect on how adoption of a growth mindset can facilitate your students' development of an academic identity.
13. What key concepts of growth mindset were most surprising to you?
14. How might you support students' cultivation of a growth mindset?

References

Atkinson, J. W. (1957). Motivational determinants of risk-taking behavior. *Psychological Review, 64*, 359–372. https://doi.org/10.1037/h0043445

Bandura, A. (1977). Self-efficacy: Toward a unifying theory of behavioral change. *Psychological Review, 84*(2), 191–215.

Bandura, A. (1986). *Social foundations of thought and action: A social cognitive theory.* Prentice-Hall.

Chamberlain, M. K. (1988). *Women in academe: Progress and prospects.* Russell Sage Foundation.

Coalition, National Native American Boarding School Healing Coalition. (2025). *US Indian Boarding School history.* https://boardingschoolhealing.org/education/us-indian-boarding-school-history/

Dweck, C. S. (2006). *Mindset: The new psychology of success.* Random House Publishing Group.

Dweck, C. S. (2017). From needs to goals and representations: Foundations for a unified theory of motivation, personality, and development. *Psychological Review, 124*(6), 689–719. https://doi.org/10.1037/rev0000082

Dweck, C. S., & Yeager, D. S. (2019). Mindsets: A view from two eras. *Perspectives on Psychological Science, 14*(3), 481–496.

Eccles, J. S., & Wigfield, A. (2020). From expectancy-value theory to situated expectancy-value theory: A developmental, social cognitive, and sociocultural perspective on motivation. *Contemporary Educational Psychology, 61.* https://doi.org/10.1016/j.cedpsych.2020.101859

Elliott-Moskwa, E. S. (2022). *The growth mindset workbook: CBT skills to help you build resilience, increase confidence & thrive through life's challenges.* New Harbinger Publications, Inc.

Ferris, S. P., & Waldron, K. (2022). Learning from senior women leaders: In their own words. In B. Cozza & C. Parnther (Eds.), *Voices from women leaders on success in higher education: Pipelines, pathways, and promotion* (pp. 81–82). Routledge.

Gale. (2025). *Women's history: Women's education in America.* https://www.gale.com/primary-sources/womens-studies/collections/womens-education#:~:text=It%20wasn't%20until%20the%20Common%20School%20Movement,time%20when%20boys%20were%20not%20in%20attendance.&text=This%20lack%20of%20access%20to%20education%20was,for%20efforts%20toward%20equal%20education%20for%20women.

Kates, A. (2017). *African Free School, first in America for Black students, found a home in Greenwich Village.* https://www.villagepreservation.org/2017/11/02/african-free-school-first-in-america-for-blacks-found-a-home-in-greenwich-village/#:~:text=The%20first%20African%20Free%20School,Rights%20and%20Social%20Justice%20Map.

Kimmerer, R. W. (2013). *Braiding sweetgrass: Indigenous wisdom, scientific knowledge, and the teachings of plants.* Milkweed Editions.

Ladson-Billings, G. (2021). *Culturally relevant pedagogy: Asking a different question.* Teachers College Press.

Love, B. L. (2019). *We want to do more than survive: Abolitionist teaching and the pursuit of educational freedom.* Beacon Press.

Malott, C. (2021). *A history of education for the many: From colonization and slavery to US imperialism.* Bloomsbury Academic.

Mejia, M. (2025). *The U.S. History of Native American boarding schools.* The Indigenous Foundation. https://www.theindigenousfoundation.org/articles/us-residential-schools

Miller, E. T., & Walker, A. V. (2023). *Antiracist pedagogy in action: Curriculum development from the field*. Rowman & Littlefield.

Milner, H. R. (2021). *Start where you are, but don't stay there: Understanding diversity, opportunity gaps, and teaching in today's classrooms*. Harvard Education Press.

Moore, E. Jr., Michael, A., & Penick-Parks, M. W. (2018). *The guide for White women who teach Black boys*. Corwin.

Morris, M. M. (2018). *Pushout: The criminalization of Black girls in schools*. The New Press.

Powell, S. D. (2012). *Your introduction to education: Explorations in teaching*. Pearson.

Ryan, R. M., & Deci, E. L. (2017). *Self-determination theory: Basic psychological needs in motivation, development, and wellness*. Guilford Publications.

Supreme Court of the United States. (1954). *Brown v. Board of Education, 347 U.S. 483 (1954) (USSC+)*. Retrieved from https://www.archives.gov/milestone-documents/brown-v-board-of-education

The National Academies of Science, Engineering, and Medicine. (2019). *Science and engineering for grades 6-12*. https://nap.nationalacademies.org/catalog/25216/science-and-engineering-for-grades-6-12-investigation-and-design

Tinto, V. (2017). Through the eyes of students. *Journal of College Student Retention: Research, Theory, & Practice, 19*(3), 254–269. https://doi.org/10.1177/1521025115621917

Weiner, B. (1985). An attributional theory of achievement motivation and emotion. *Psychology Review, 92*, 548–573.

Weiner, B., Van Lange, P., Kruglanski, A., & Higgins, E. (2012). An attribution theory of motivation. In B. Moulding, N. Songer, & K. Brenner (Eds.), *Handbook of theories of social psychology* (Vol. 1, pp. 135–155). Sage UK.

Wells, L. M. (2024). *There are no deficits here: Disrupting anti-Blackness in education*. Corwin.

Wigfield, A., & Eccles, J. S. (2000). Expectancy-value theory of achievement motivation. *Contemporary Educational Psychology, 25*, 68–81.

Zachary, L. J., & Fain, L. Z. (2022). *The mentor's guide: Facilitating effective learning relationships*. Jossey-Bass.

Zeldin, A. L., Britner, S. L., & Pajares, F. (2008). A comparative study of the self-efficacy beliefs of successful men and women in mathematics, science, and technology careers. *Journal of Research in Science Teaching, 45*(9), 1036–1058.

Zeldin, A. L., & Pajares, F. (2000). Against the odds: Self-efficacy beliefs of women in mathematical, scientific, and technological careers. *American Educational Research Journal, 37*(1), 215–246.

Zhou, L. (2021). Cross-cultural mentoring: Cultural awareness & identity empowerment. *InterActions: UCLA Journal of Education and Information Studies, 17*(1). https://doi.org/http://dx.doi.org/10.5070/D417150292

PART II
Facilitating the Peer Mentoring Relationship

4
GETTING STARTED

Introduction

Now that you've made it to Part II of this book, you should have a relatively strong foundation for understanding the basic constructs that support peer mentoring. You might be wondering now, what do I do? How do I get started? What does all of this mean in the context of actually supporting my students? Well, this is the chapter where we get into the nitty gritty of what to do. So, let's get started.

Defining Peer Mentoring

As a reminder, peer mentoring can be defined as "a mentoring relationship that occurs between two or more peers, where one peer is usually more experienced than the other(s) and is referred to as the *mentor*. The less experienced peer(s) is referred to as the *mentee*" (Rockinson-Szapkiw et al., 2020, p. 2). Within this relationship, it is not uncommon for the mentor and mentee to also be considered near-peers—that is, the mentor is slightly more advanced in their learning progression (or training) than the mentee (Akinla et al., 2018). That's okay. For the purposes of this book, peer mentoring is considered mentoring that is occurring between two or more students with the understanding that they will not necessarily be at the same level in their respective degree programs or even have the same experience level. This will inevitably include near-peer mentoring.

Within the peer mentoring relationship, you as the faculty member will not necessarily be engaging in the mentoring activities to the same extent

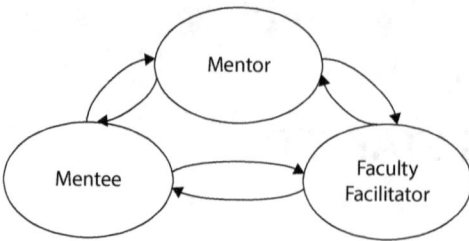

FIGURE 4.1 The peer mentoring relationship is a collaborative, reciprocal relationship.

as the mentor will. That is, you are meant to support and facilitate the mentor and mentee as they work together to engage in mentoring. Again, you are a support—a scaffold, providing guidance and ensuring that the peer mentoring relationship stays on track. Think of yourself as the "guide on the side." With this in mind, let's discuss the various roles within the peer mentoring relationship.

Peer Mentoring Roles

Keeping in mind that the peer mentoring relationship is a reciprocal, collaborative relationship (see Figure 4.1), there are general expectations for the roles of each individual within the peer mentoring relationship. While these role expectations are flexible and will vary depending on need, context, and other external factors, they can serve as a basic guideline so that everyone understands what they are meant to contribute to the relationship, when, and how.

The Role of the Peer Mentor

The mentor within the peer mentoring relationship should be the individual (the student) that has the most senior level of experience or training. This could mean that the mentor is more advanced in their degree program or perhaps has a greater level of field experience or outside training. Again, there are no hard and fast rules here. This is merely a general guideline. The peer mentor should be someone who is organized, can manage time well, who possesses a strong level of intrinsic motivation, and can initiate steps to move the relationship forward autonomously. The mentor will be responsible for ensuring that the peer mentoring relationship operates efficiently and that everyone is making progress toward meeting their goals. The mentor will be responsible for initiating contact with their mentee(s), establishing shared expectations within the relationship, and assisting the

TABLE 4.1 A List of the characteristics that mentors should possess as reported by mentees

Mentor Characteristics

- Organized
- Manages Time Well
- Dedicates Appropriate Time to the Mentoring Relationship
- Initiates Communication
- Communicates Frequently and Clearly
- Honest and Trustworthy
- Dependable
- Accountable
- Safe and Inclusive
- Empathetic and Understanding
- Recognizes and Acknowledges Growth
- Knowledgeable and Resourceful
- Shares Resources and Networking Opportunities
- Motivated and Motivating
- Positive and Encouraging
- Genuine
- A Role Model

mentee in determining a purpose for the relationship, setting goals, and demonstrating accountability. The mentor will serve as a role model but will also be in a position to learn and grow!

I have had the privilege of working with community college, undergraduate, graduate, and doctoral STEM students as they have engaged in peer mentoring relationships within a variety of contexts. As a result, I've been able to review feedback from mentors and mentees on their experiences. The characteristics shown in Table 4.1 are what mentees have reported are the characteristics that effective mentors possess based on my own research as well as the available research literature (Jones & Wendt, 2025; Rockinson-Szapkiw et al., 2020; Rolfe, 2021; Wendt & Jones, 2024b; Wendt et al., 2021). The characteristics that mentees report are essential for mentors to effectively engage in a peer mentoring relationship include being organized; being able to manage time well; dedicating appropriate time to the mentoring relationship; initiating communication with the mentee(s); communicating clearly and frequently; being honest and trustworthy; being dependable; possessing a sense of accountability; being perceived as safe and inclusive; being perceived as empathetic and understanding; recognizing and acknowledging mentees' growth; possessing content knowledge and knowledge of resource availability; being willing to share resources and professional networks; personally motivated; being motivating, positive, and encouraging; being genuine; and serving as a role model.

For instance, in my own research, a mentee confided that the most critical step of the peer mentoring relationship was when the mentor made initial contact. This set in motion the trajectory of the entire relationship. "You know...reaching out and asking for times to meet," noted the mentee. Coupled with frequent check-ins and clear, concise communication, this helped the mentee to be able to gain a level of comfort with their mentor that allowed them to feel able to depend on their mentor. Another mentee shared with me that an essential characteristic of an effective peer mentor was being someone that the mentee could lean on—someone they could reach out to, that they could trust, that they could feel comfortable with; "Knowing that I had a mentor who I can call and who I can actually express myself to without any judgment," the mentee shared. Importantly, mentees have frequently shared with me the importance of their mentor not only being motivated but also *motivating*. That is, they must have a certain level of personal motivation but also be able to encourage their mentees. One mentee shared about her mentor, "The main thing she always told us was to follow our hearts and follow our passions" (see Figure 4.2). This gave the mentee a sense of being seen as an individual with dreams and ability,

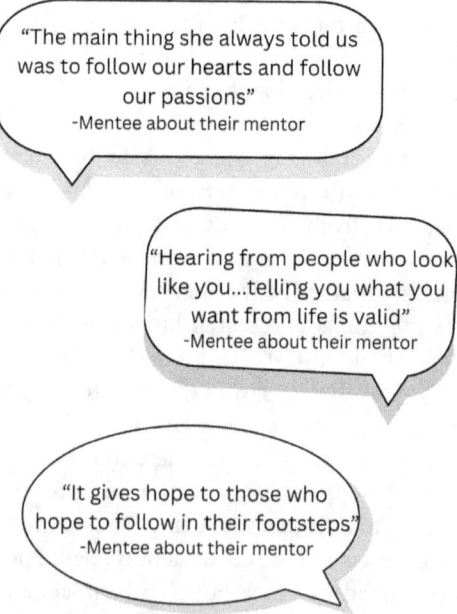

FIGURE 4.2 Quotes from actual mentees describing the characteristics of effective mentors.

which further enabled the mentee to feel confident in her goals. In a focus group that I held, a mentee shared that her mentor was consistent in recognizing the challenges that they each faced while simultaneously reinforcing that the mentee was capable. The mentee said, "That was a message that I was getting from her saying, you know what? She's right. I can do it. It's not hard. Well, *this* is hard, but I can do it."

Overall, mentees perceived good mentors as being someone whom they could look to as a role model—even if they had similar experiences and levels of training and expertise (such as near-peers). Mentees wanted mentors whom they could trust to truly understand mentees' experiences, to demonstrate empathy, and to provide the encouragement necessary for the mentee to persist. As shared in a focus group, one mentee said, "Hearing from people who look like you or who have experienced similar things as you. Telling you that what you want from life is valid. You are deserving of a place at whatever table you have in mind. That's so important to hear." Another mentee chimed in, "It gives hope to those who hope to follow in their [mentor's] experience."

The Role of the Peer Mentee

The mentee is the individual(s) (or student(s)) within the peer mentoring relationship that holds the more junior level of experience or training. The mentee should have a sincere desire to be mentored and should be open to communication and feedback with their mentor. They should be open to sharing in an honest and transparent manner and trying new experiences. The mentee should possess a strong level of intrinsic motivation. They must be willing to "do the work" and not rely on their mentor to tell them what, when, and how to do things. Rather, they should be able to self-initiate, which includes being able to reflect on what they need and when as well as be able to communicate those needs clearly and in a timely manner to their mentor. The mentee should be willing to share their experiences and expertise with the mentor as the peer mentoring relationship is indeed collaborative and reciprocal. Importantly, trust and transparency are key (APA, 2012). In Table 4.2, you will find a list of characteristics that mentors have reported that effective mentees should possess in my own research as well as in the research literature (Rockinson-Szapkiw et al., 2020; Rolfe, 2021; Wendt & Jones, 2024a). The characteristics that mentors report are essential for mentees to effectively engage in a peer mentoring relationship include an openness and willingness to be mentored; dedicating appropriate time to the mentoring relationship; responding to communication from mentors in a timely manner; communicating clearly and frequently; asking

TABLE 4.2 A list of the characteristics that mentees should possess as reported by mentors

Mentee Characteristics
• Willing to Be Mentored • Dedicates Appropriate Time to the Mentoring Relationship • Responds to Communications Frequently and Clearly • Asks for Help When Needed • Honest and Trustworthy • Dependable • Patient • Willing to Learn and Try New Approaches • Motivated • Open to Reflection and Constructive Feedback • A Self-Starter

for help when needed; being honest and trustworthy; being dependable; being patient with mentors while understanding that they have other responsibilities that require their time and attention; being willing to learn and try new approaches; being motivated; being open to self-reflection and constructive feedback; and being a self-starter.

For instance, several mentors shared in one of the focus groups that I held that one of the most important characteristics of a "good" mentee is "just being open to talking and open to sharing"—just "being open and being honest about everything" (see Figure 4.3). This was important to the mentors as it allowed them to "take a step back and help"—to provide targeted support that the mentee genuinely needed rather than needing to guess or make assumptions about what the mentee needed. In line with openness and honesty, the mentors in one of the focus groups concurred that active participation and engagement on the part of the mentee is necessary for an effective peer mentoring relationship. One mentor shared that "participation goes a long way." Another mentor shared in an interview that mentees "can't build a relationship if you are not present. Showing up. That's half the battle."

Patience was another key characteristic, particularly as it aligns with understanding the mentors' humanity. One mentor shared in a focus group that they appreciated the mentee being understanding of their own learning process and the reciprocity of the mentoring relationship so that the mentor didn't need to "pretend to be a professional." In one interview with a female mentor, I was told that an effective mentee was one who could "just be present and be adapting and be open and be understanding. Be kind. These things I think matter."

FIGURE 4.3 Quotes from actual mentors describing the characteristics of effective mentees.

The Role of the Faculty Facilitator

The role of the faculty facilitator is a unique role that, in my opinion, is often underutilized and, in many ways, minimized—usually by faculty themselves. Oftentimes, faculty view the student-to-student peer mentoring relationship as one that can effectively run on its own with minimal faculty input and involvement. However, that assumption is not always accurate. In fact, in my work in peer mentoring, I have found that students actually desire *increased* faculty involvement in their peer mentoring relationships. They don't want to be matched with their respective mentors and mentees and left to their own devices to figure things out. They actually desire to have faculty involved as one form of accountability partner. There is a fine line, though. Faculty should be involved in the relationship to an extent, but should also afford the mentor and mentee autonomy. This is critical for maintaining a space within which mentor and mentee can freely share their feelings and needs and can develop essential skills such as critical thinking and problem-solving. There is great benefit to mentors and mentees, especially when they are near-peers, to being able to function within the peer mentoring relationship without perceptions of faculty judgment. After all, as the faculty facilitator, you aren't going to be able to be present for your students forever. There will be some point where they will need to "fly from the nest" and do things on their own. Thus, as a facilitator, your role is to prepare them for that time. With this in mind, the role of the faculty facilitator is exactly that—a facilitator, a "guide on the side."

Determining the Peer Mentoring Groups

The faculty facilitator is responsible for matching the mentor and mentee(s). That is, the faculty facilitator must consider various factors in order to appropriately determine who will work best in mentor roles, who will work best in mentee roles, and which combination of mentor and mentee will be most appropriate within various contexts. Research studies support that there is great benefit to ensuring "fit" or match between mentor and mentee (Mondisa, 2018; Saffie-Robertson, 2020), especially when considering race, ethnicity, or culture. When a mentee can observe their mentor and see someone that has characteristics like them, they are more likely to adopt the mindset that they, too, can be successful at specific tasks. An appropriate match helps mentees visualize themselves in the position that the mentor is in. That is, they can *see* themselves being successful by seeing successful others.

There is also benefit to matching mentor and mentee by interest, expertise, content area, or goal. For instance, in my work in peer mentoring, one mentee reiterated the importance of faculty taking "into account the dynamic of the mentee and mentor to better suit both the mentee and the mentor, so they could both get something out of it" during a focus group conversation. Another mentee in a different focus group confirmed that "It was good to see people who look like me. It gave me a sense of belonging because I could identify with each of their stories."

This does not necessarily mean, however, that the mentor and mentee must possess characteristics that match entirely. For instance, in my work in peer mentoring that primarily focused on supporting women in peer mentoring relationships, male mentees reported benefits from observing and engaging with successful women in STEM fields—even when gender identity, racial identity, and disciplinary area did not necessarily match (Wendt & Jones, 2024b). That being said, there is also a benefit to having diversity within the peer mentoring relationship. Identities are malleable and can morph and change over time (Miller, 2021). Likewise, the intersection of identities

TABLE 4.3 Questions to ask mentors and mentees to help determine match or fit

Questions to Ask for Fit and Match

What is your degree program and concentration?
In what year are you within your degree program (e.g., freshman, sophomore, junior, senior, graduate student, post-graduate student)
What is your prior experience in your field of interest?
What do you hope to gain from the peer mentoring experience?
What strengths do you have to share through the peer mentoring experience?
Are there any characteristics that you hope for your mentor/mentee to have?
Are there any personality quirks that you would prefer to avoid with a peer mentor/mentee placement?
On what days and times would you be available and willing to engage in the peer mentoring experience?
Are there any other specific needs or preferences that you feel it would be important to share?

---

introduces unique combinations that, no matter how hard we try, cannot be perfectly replicated or matched (Taylor, 2017)—nor should they. However, finding some sort of alignment is imperative so that the overall purpose of the peer mentoring relationship can be fulfilled and so that both mentor and mentee will be able to meet their desired goals. Having students fill out a quick questionnaire or even having a short conversation with potential mentors and mentees will help yield insights on what characteristics are most important to your particular students. What works for one peer mentoring relationship may not necessarily work for the next peer mentoring relationship, so consider context, need, and the unique attributes of your students. Table 4.3 presents some questions that you may want to ask potential mentors and mentees to determine an appropriate match or fit.

Peer Mentor and Peer Mentee Training

The faculty facilitator is also responsible for ensuring that the mentor and mentee are trained appropriately. One of the biggest mistakes that occurs within mentoring in general is that the assumption is made that just because an individual is more advanced in their training or skill that they automatically know how to mentor (McGee, 2016; Pfund et al., 2016; Rockinson-Szapkiw et al., 2020; Wendt & Jones, 2024a). This is such a misguided view, especially given the massive impact—both positive and negative—that mentoring can potentially yield on an individual. Mentoring, when implemented correctly, can yield wonderful benefits! On the other hand, mentoring, when implemented poorly, can also be detrimental. Obviously, if you are reading this book, you are interested in the peer mentoring relationship being a productive and positive one. Thus, make sure your mentors and mentees are trained.

There are various resources that can assist with providing mentors and mentees with peer mentoring training (see Chapter 7 for suggestions). Within the context of peer mentoring among STEM students in particular, through my collaborative work, a series of online training modules for peer mentors and mentees has been developed, tested, and published for use. You can find these training modules on the eSTEM Mentoring website (www.eSTEMmentoring.com) (eSTEM Mentoring, 2025). I would strongly encourage you to have mentors and mentees complete the training in tandem with reading and completing the workbook *Navigating the Peer Mentoring Relationship: A Handbook for Women and Other Underrepresented Populations in STEM* (Rockinson-Szapkiw et al., 2020). (Note that while I do understand that this workbook is geared toward women and other underrepresented populations, it has been used by men with good success and is still highly relevant.)

Facilitating the Phases of Peer Mentoring

Once your students are adequately trained and appropriately matched, then they will be ready to move through the four phases of peer mentoring (which are described a little later in this chapter in much more detail). The faculty facilitator will be responsible for holding the mentor and mentee accountable throughout these four phases, facilitating mentor and mentee movement through the phases, and will serve as a secondary resource when the mentor or mentee needs additional direction or sources and/or challenges arise that could benefit from an unbiased third party to mitigate or resolve. During the four phases of peer mentoring, the faculty facilitator should truly adopt the "guide on the side" approach. That is, let the mentor and mentee work together to drive the peer mentoring relationship. As a faculty facilitator, you are there to assist as needed and only when needed. The amount of facilitation needed will fluctuate depending on each individual peer mentoring relationship and may also fluctuate throughout the duration of the relationship. Frequent and clear communication with your students will be essential during the four phases of peer mentoring.

Keep in mind, though, that it is not the job of the faculty facilitator to provide all of the answers to the mentor and mentee (nor is it the job of the mentor to provide all of the answers to the mentee). Rather, the faculty facilitator should support the mentor and mentee in using critical reflection in order to gain insight that will enable them to take action toward meeting their goals (Rolfe, 2021) and in providing additional assistance as needed. This is key to ensuring that you as the faculty facilitator do not over-involve yourself in the peer mentoring relationship both from the standpoint of providing a certain level of privacy and autonomy to the mentor and mentee and in holding boundaries that protect your use of time and availability (remember, we are trying *not* to overburden faculty!). We'll dive deeper into this in Chapter 5.

Sharing Opportunities and Networks

Think back to when you were an undergraduate or graduate student. How connected were you to professional organizations? How connected were you to a professional network? Did you know anyone who worked in your field of study? How did you first get your foot in the door?

Each of us relies on others as we learn, grow, and move through our professional journey. There is not a single one of us who did absolutely everything on our own. We have relied on someone at some point to give us direction, share a resource, share a contact, or crack a door of opportunity open—if even to a very small extent. As part of the faculty facilitator responsibilities, you will be responsible for sharing opportunities (such as internships, conferences, research opportunities, etc.) as they arise, and will be responsible for connecting mentors and mentees to your professional networks as appropriate. This could mean sharing with your students a list of professional organizations that have been most helpful to you, sharing an internship opportunity that landed in your email inbox, or even finding a way to collaborate with students on a research project, paper, or conference presentation. Again, this will all differ depending on the particular needs of your students. But do not minimize how impactful your actions can be in helping your students' progress through their own journey. (We'll discuss this in more detail in Chapter 6.)

Serving as a Role Model

Finally, the faculty facilitator is responsible for serving as a role model for both the mentor and mentee. Thus, it is essential that the faculty facilitator align their words and actions with their values, demonstrating kindness, respect, responsibility, and integrity (refer back to Chapter 3 if needed). How you communicate with your students is critical! Let's be honest—many faculty, especially those who adopt a more traditionally minded approach—tend to believe that with all of their knowledge and degrees, they should be highly revered and esteemed. These faculty believe themselves to be the "end all, be all," walking around campus as if they were royalty or gods. While we can all agree that faculty are knowledgeable and should be respected, I'm hopeful that we can also agree that faculty who adopt a more welcoming approach are more likely to impact students in a positive manner—especially those students who desire individual guidance, conversation, and a more personalized professional relationship. Thus, talking to your students like human beings—treating them with respect and dignity—is absolutely essential. Kindness, humility, and expertise can all exist simultaneously! And this, of course, ties directly to the use of culturally responsive practices that were discussed in Chapter 3.

Consider what one mentee shared with me at the conclusion of their peer mentoring program when reflecting on when he first met me, "I remembered the first meeting when I came to pick up the computer and you told me 'I have a biology background' when I told you that I'm a biology major. So it says here [referring to the degrees displayed on the wall] and she's a doctor and she's talking, like, normal. You were like an example and that's exactly what we need." This particular mentee shared on several occasions how appreciative he was that I provided a warm, welcoming space for students—a place where they could share challenges and successes without judgment. A place where they could just *be*. Be sure to consider how students perceive you as a faculty member and take steps to ensure that you are creating an environment and persona that encourages students to connect with you as a resource. In summary, Figure 4.4 presents a list of faculty facilitator responsibilities.

The Faculty Facilitator is Responsible For:

- Matching the mentor and mentee(s)
- Providing training
- Serving as an accountability partner
- Providing resources as necessary
- Sharing opportunities
- Connecting the mentor and mentee to the faculty facilitator's professional network
- Assisting in mitigating challenges as necessary
- Serving as a role model

FIGURE 4.4 Faculty facilitator responsibilities when supporting the peer mentoring relationship.

Now that you have a basic understanding of the roles of the peer mentor, peer mentee, and faculty facilitator, let's explore the phases of peer mentoring so that I can then explain how the faculty facilitator can *actually engage in* facilitating the peer mentoring relationship.

The Phases of Peer Mentoring: Doing the Actual Facilitating

As with any relationship, the peer mentoring relationship will journey through several phases. There are four main phases of the peer mentoring relationship that the mentor and mentee will journey through: (1) relationship building, (2) relationship negotiation and expectation setting, (3) relationship working, and (4) relationship termination (Rockinson-Szapkiw et al., 2020) (see Figure 4.5). In the relationship building phase, the mentor and mentee begin getting to know one another by developing trust and rapport (APA, 2012). Key to this phase is the practice of active and empathic listening. This also requires both the mentor and mentee to openly share their feelings, concerns, and aspirations. This is where alignment of words and actions with values is most important as this is the phase where the mentor and mentee determine whether they are sufficiently comfortable with one another to begin building a relationship of transparency and trust (remember Chapter 3!). Most often, this phase begins the moment that the mentor and mentee are first introduced to one another.

In the relationship negotiation and expectation setting phase, the mentor and mentee work together to determine an overall shared purpose for their mentoring relationship. Additionally, they work together to develop individual goals, set expectations, and establish healthy boundaries that will guide their relationship. This phase typically occurs immediately after the relationship building phase, but should be revisited as often as necessary as the relationship progresses and needs to change.

In the relationship working phase, the mentor and mentee take action to work toward meeting the individual goals that they have collaboratively developed. This is the phase during which the mentor provides the most guidance, assisting the mentee by providing information, resources, and suggestions for action, and by providing constructive feedback. During this phase, both the mentor and mentee engage in reciprocal sharing of information for knowledge building and networking while simultaneously working together to support, encourage, and affirm one another. Importantly, this phase requires active participation and commitment on both the part of the mentor and the mentee.

In the relationship termination phase, the mentor and mentee make the decision to end the relationship. Keep in mind that ending the relationship is often a natural part of learning and growth and does not necessarily

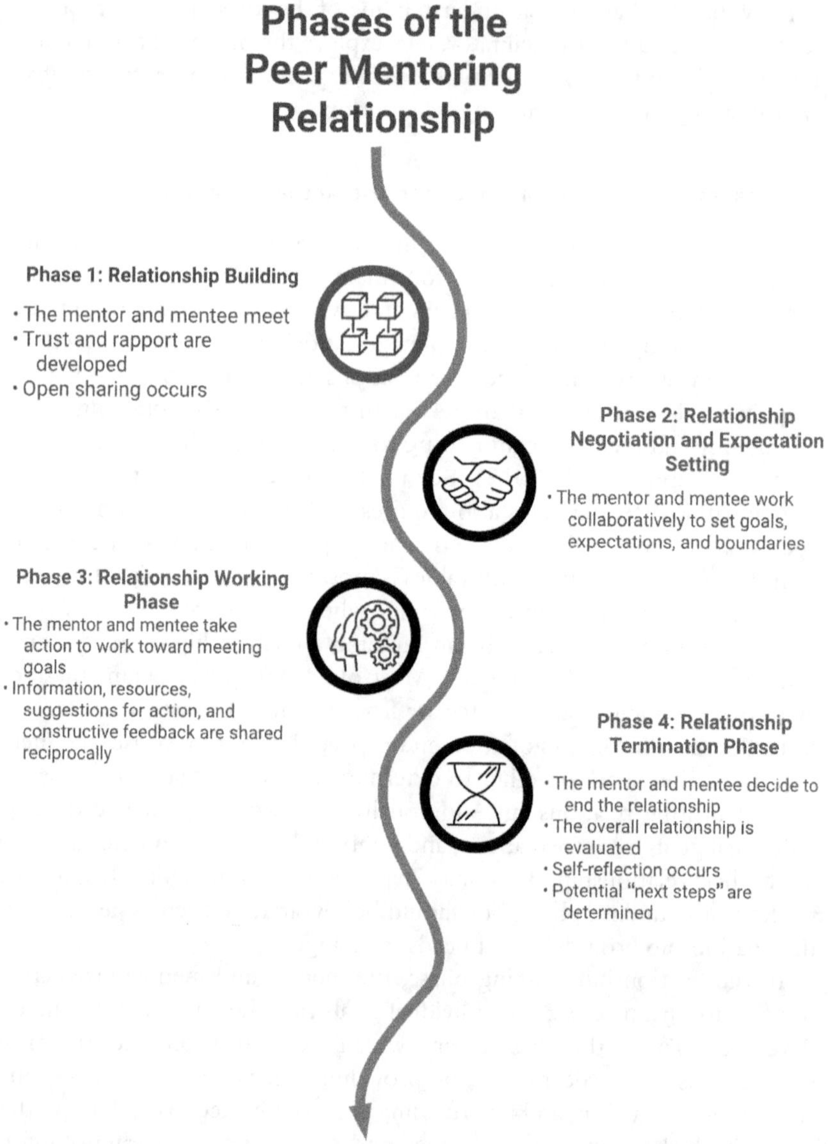

Phases of the Peer Mentoring Relationship

Phase 1: Relationship Building
- The mentor and mentee meet
- Trust and rapport are developed
- Open sharing occurs

Phase 2: Relationship Negotiation and Expectation Setting
- The mentor and mentee work collaboratively to set goals, expectations, and boundaries

Phase 3: Relationship Working Phase
- The mentor and mentee take action to work toward meeting goals
- Information, resources, suggestions for action, and constructive feedback are shared reciprocally

Phase 4: Relationship Termination Phase
- The mentor and mentee decide to end the relationship
- The overall relationship is evaluated
- Self-reflection occurs
- Potential "next steps" are determined

FIGURE 4.5 The phases of the peer mentoring relationship.

Sources: From *Hand, finger, handshake icon*, by Puckung graphic design factory, Iconfinder (https://www.iconfinder.com/icons/7417587/hand_finger_handshake_hold_greeting_icon). CC BY 3.0 and from *Essential, hourglass, web icon*, by Vinzence Studio, Iconfinder (https://www.iconfinder.com/icons/3401850/essential_hourglass_web_icon). CC BY 4.0.

indicate anything negative about the relationship! For instance, perhaps the mentor is graduating and moving on to a new beginning in a new career. Or perhaps the goals of the peer mentoring relationship have been met, and the relationship has come to its natural end. Occasionally, though, the mentor and mentee may find that they are no longer a good fit for one another or that the peer mentoring relationship is not working to support their achievement of goals. In this case, the mentor and mentee may decide that it is best not to continue the peer mentoring relationship. This, too, is okay and is valuable information that should inform future "next steps." No matter the reason for ending the relationship, this is the phase where the mentor and mentee evaluate the overall outcomes of the peer mentoring relationship, engage in assessing whether goals were met, self-reflecting on what worked well and what could use improvement, and determining "next steps" (if any) for the relationship. This is also the phase where the mentor and mentee should celebrate their accomplishments during the time that they have worked together—even small accomplishments!

In the following sections, I'll focus on the faculty facilitator role as the peer mentoring relationship progresses through each of the phases.

Facilitating Match and "Fit"

As mentioned earlier in this chapter, one of the responsibilities of the faculty facilitator is to match or "pair up" the mentor and mentee, ensuring some sort of alignment between demographic characteristics, experience, needs, and/or goals—all of which will be dependent upon the particular context within which the peer mentoring relationship will be situated as well as the unique individuals that will be engaging in the peer mentoring relationship. It is so very important to remember that what is important to one individual when it comes to match or fit may not be important to the next individual, so do not make sweeping assumptions. Rather, this is where the skills of active and empathic listening are imperative. What is it that is important to your particular students? What is it that will best enable them to share openly with each other, set goals, and take action to meet those goals successfully? You can begin to determine the answers to these questions by simply asking your students and, of course, listening to their responses. This can be done through a simple questionnaire or through a short individual conversation. I personally would recommend an individual conversation as this will also allow you the opportunity to get to know each mentor and mentee as their own unique person. This will also afford you the opportunity to experience and interpret subtle nuances in responses, including nonverbal cues (or body language).

In addition to the questions shared in Table 4.3, some additional questions that you might want to ask—especially if you are able to have a conversation with potential peer mentors and mentees– include the following:

- Have you ever been in a mentoring relationship before? If so, what was your experience? What worked well? What didn't work well? What would you do differently if engaging in the mentoring experience again?
- What are your personal goals for the peer mentoring experience? (This question should focus on the individual's goals for themselves—not necessarily for the peer mentoring partnership.)
- What unique experiences do you bring to the peer mentoring relationship?
- What types of experiences, expertise, or knowledge would you like to see in your mentor or mentee? (This question should help get to the heart of whether experience, training, or discipline area will be most integral to ensuring match or fit.)
- With what type of person or people do you find yourself most comfortable? (This question should help get to heart of what personal characteristics are most integral to ensuring match or fit. This will also help attend to personality traits that could enhance or potentially inhibit the peer mentoring relationship.)
- What are your biggest concerns for the peer mentoring relationship—especially with whom you will be potentially partnered with?

While these questions are not the only questions that can be asked, they should serve as a good start to getting to know your students and determining which attributes will be most important when matching mentors and mentees.

Once you have matched mentors and mentees, keep in mind that your work is not complete! It is very important to monitor whether the match is indeed appropriate for each student's needs, whether personalities mesh, and whether goals align. Here again, I encourage active and empathic listening—check in with your students individually to see how the relationship is going. What is working with the match? What is not working with the match? Does the match need to be reconsidered? If so, this does not mean that anyone has done anything wrong or that you are terrible at matching students. There are myriad factors that can influence whether a match works and whether it doesn't, and these factors are out of your control. Thus, listen to your students, adjust, and carry on. (See Chapter 7 for a link to the University of Wisconsin mentoring page, which includes an Assessing Fit Checklist that can be used to query your students and measure the feasibility of the matches that you have created.)

Facilitating Relationship Establishment (Phase 1)

Once you have gotten to know your students and thoughtfully matched mentors and mentees, it is essential that the mentors and mentees are provided with adequate training (as discussed previously). Once training is completed and your students have gained the appropriate mentoring skills and competencies to effectively engage in a peer mentoring relationship, they will then begin moving through the four phases of the peer mentoring relationship. If you recall, the first phase is the relationship building phase. This is where the faculty facilitator should assist the mentor and mentee in being introduced to one another. The faculty facilitator does not necessarily need to be present to do so—an email introduction could certainly suffice. However, consider context, need, and the purpose of your particular peer mentoring program in order to make the best judgment about how the introduction should occur, keeping in mind that in-person introductions are more personal and can afford individuals the opportunity to experience non-verbal cues (such as body language) that can often assist in establishing relationships more effectively and more efficiently. You will want to introduce yourself as well as you will be an integral part of the peer mentoring relationship as the faculty facilitator. Acquaintance activities can be helpful in aiding this process, such as icebreakers, name games, and so on. There are many, many resources to help generate ideas if needed (one simply needs to do a quick internet search for "Acquaintance Activities"). However, several recommended resources can be found in Chapter 7—all of which can be easily adapted to meet the particular needs of your students within your particular context of peer mentoring.

Facilitating Purpose, Goals, and Boundaries (Phase 2)

During the relationship negotiation and expectation setting phase, the faculty facilitator is responsible for assisting the mentor and mentee in determining the overall purpose for the peer mentoring relationship, setting individual goals that are feasible, measurable, and achievable, and establishing healthy boundaries that will support an effective peer mentoring relationship. During this phase, the faculty facilitator should clearly communicate any expectations that they have of the mentor and mentee during the peer mentoring relationship, which might include frequency of interactions (between mentor and mentee, between mentor and faculty, and between mentee and faculty, for instance); duration of interactions (as well as the overall duration of the peer mentoring relationship, if relevant); requirements for meeting notes, formal mentoring plans, or written goals (as relevant); and confidentiality to ensure trust and adhere to ethical responsibilities (see Chapter 5). Likewise, this phase is where the mentor and mentee will also work together to establish their own expectations for each other, including the creation of a mentoring agreement.

Purpose and Goals

When assisting the mentor and mentee in determining the purpose of the overall peer mentoring relationship, it is helpful to determine the main focus of the relationship. That is, will the main focus be on developing academic knowledge or skills? Will the main focus be on developing career and professional knowledge or skills? Or will the main focus be on developing personal and psychosocial knowledge or skills? Neither focus is more or less important than another and, oftentimes, the peer mentoring relationship will attend to more than one specific area of focus. But determining the main "why" is essential for guiding your students through an effective peer mentoring relationship as it sets the tone for the entire relationship.

Once the purpose of the peer mentoring relationship is collaboratively decided on, the mentor and mentee will work to develop individual goals that are feasible, measurable, and achievable. Most of us are probably familiar with SMART goals as they have been ubiquitous within the workforce and professional development. SMART goals are developed and written in a manner so that they are a) specific, b) measurable, c) achievable, d) realistic, and e) time sensitive. In this case, the mentor and mentee may need assistance in making sure that their goals are indeed SMART (see Figure 4.6).

Boundaries

Few people, in my experience, are naturally gifted at setting healthy boundaries that support an appropriate balance in attending to the needs of self and others. While socioemotional skills development (which

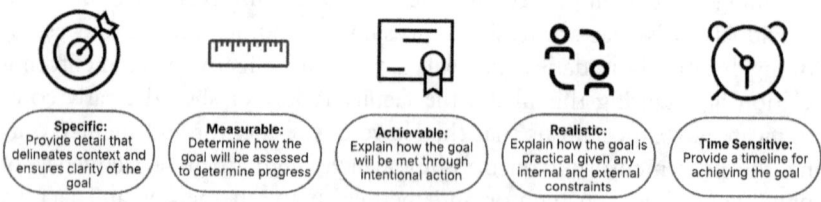

FIGURE 4.6 A visualization of SMART goals.

Source: From *Collaborate icon*, by Carbon Design, Iconfinder (https://www.iconfinder.com/icons/9044227/collaborate_icon/). CC BY 4.0.

includes setting boundaries and maintaining boundaries) has become more prevalent in primary and secondary schools, many of us still find this task a bit challenging. Thus, your students may need assistance in determining what boundaries will best protect their well-being and will allow them to be fully present and responsive when in the peer mentoring relationship. Boundaries might include when and how to contact one another, what types of resources and assistance one is willing to provide and under which particular circumstances, topics that are appropriate as well as those that may be triggering or are generally not welcomed within the relationship (if any), under what specific circumstances information can be shared outside of the peer mentoring relationship (in order to maintain confidentiality), among others that will really be dependent on individual and context. You will also want to communicate any boundaries that you have with the mentor and mentee. For instance, should the mentor and mentee contact you outside of office hours? If so, until what time? How? Is it appropriate for them to simply stop by, or is an appointment necessary? If there is a question, are there particular steps that the mentor and mentee should complete prior to reaching out to you? If there is a conflict, what actions do you expect to be completed prior to your involvement? Your boundaries should align appropriately with the expectations already established as well as the amount of access, interaction, and oversight that will be necessary in order for the mentor and mentee to meet their respective goals. Boundaries will also serve to protect your workload and help ensure that your involvement in the peer mentoring relationship is not such that it is overburdening or overwhelming. (A more robust discussion on the importance of boundaries is provided in Chapter 5.)

Facilitating the Creation of a Mentoring Agreement

Once the mentor and mentee have agreed upon the expectations of the peer mentoring relationship, with assistance from you if needed, it is best practice for the mentor and mentee to create and sign a mentoring agreement. You can think of the mentoring agreement as a contract of sorts that outlines and articulates clearly everyone's responsibilities in the relationship. If you wish to be part of this agreement—that is, outline your part within the peer mentoring relationship in writing—as well, you can. However, the mentoring agreement should primarily focus on the relationship between the mentor and mentee and should be drafted by the mentor and mentee collaboratively (Rockinson-Szapkiw et al., 2020; Zachary & Fain, 2022).

In the mentoring agreement, the mentor and mentee should collaboratively negotiate and decide on objectives for the relationship, goals (both for the relationship and individually), measures for success, individual responsibilities, accountability measures, time expectations and constraints, and a general mentoring plan (Rockinson-Szapkiw et al., 2020; Zachary & Fain, 2022). The overarching idea is to ensure that the peer mentoring relationship is poised to be effective for all parties involved. (For resources to support the various facets of constructing a mentoring agreement, including examples, see the National Academies of Sciences, Engineering, and Mathematics website listed in Chapter 7 under Mentoring Training, Tools, and Guides as well as Chapter 5 of Zachary and Fain's (2022) *The Mentor's Guide: Facilitating Effective Learning Relationships*.)

Supporting the Work of Peer Mentoring (Phase 3)

During the relationship working phase, the faculty facilitator primarily serves as the "guide on the side." During this phase, the mentor and mentee will be working together to meet the goals that they have collaboratively set using the mentoring plan that they have collaboratively negotiated. The faculty facilitator should, during this phase, provide whatever guidance is necessary to keep the mentoring relationship on track toward meeting goals and aligning with expectations. As noted previously, active listening is essential as your cue to become actively involved as the faculty facilitator should come from your students. That is, be actively involved when needed—based on real-time mentor and mentee needs. The rest of the time, your involvement should be more passive. Remember, the faculty facilitator is the "guide on the side," poised to provide support to the mentor and mentee as and only when needed.

During this phase, I like to adopt Zachary and Fain's (2022) perspective on growth through the mentoring process. Zachary and Fain's perspective of mentoring is grounded in the three core conditions for facilitating learning (Daloz, 1999): *support, challenge, and vision*. Through *support*, the mentoring relationship is managed so that an environment conducive to learning is established (Zachary & Fain, 2022). Through *challenge*, the momentum of the mentoring relationship is preserved through providing opportunities for challenge, process monitoring, and progress evaluation. Through *vision*, the movement of the mentoring relationship is ensured through working toward a shared vision, engaging in self-reflection, and self-assessment. The faculty facilitator's role during the relationship working phase, if it could be summed up succinctly, is to enable growth through the mentoring relationship. The faculty facilitator does this through support, challenge, and vision (Figure 4.7).

FIGURE 4.7 The three core conditions for facilitating learning.

Source: From *Customer, user, userphoto icon*, by Artyom Khamitov, Iconfinder (https://www.iconfinder.com/icons/1564535/customer_user_userphoto_account_person_icon). CC BY 3.0.

Support: Sharing Opportunities for Learning and Growth

During the relationship working phase, you will *support* your students' growth through the mentoring relationship. One way that you can support growth is by sharing opportunities with your students, such as internship opportunities, professional development opportunities, and research opportunities to name a few. This is the part of the mentoring relationship where you can make a substantial impact on students by sharing access to your professional networks and by more generally opening doors for students so that they can enjoy experiences that otherwise they may not be privy to. During this phase, I would strongly encourage you to think outside the box. What activities are you currently engaged in that you could invite students to participate in, giving them valuable experience and expanding their opportunities? For instance, are you writing a paper that you could invite a student to collaborate on, giving them valuable writing and publishing experience? Are you presenting a poster that a student could assist with that would provide them with experience in disseminating results or, critically, experience attending a professional conference? Are you working on a research project that could include a student—even in some small way—that might open doors to both opportunity and interest in your field? Consider where you have existing opportunities that could be adapted to best support your students as well as new opportunities that will enable you to learn and grow as well. Where can you branch out and try different activities in a new way that includes students? This is also a great time to introduce your students to professional organizations that can facilitate their learning and growth both as students and as professionals within their future careers. (We'll discuss how to facilitate opportunities further in Chapter 6.)

Challenge: Momentum and Problem Solving

During the relationship working phase, you will also work with your students through *challenge*. The primary method through which you will do this is by monitoring, encouraging, and facilitating students' motivation and momentum as they work toward meeting their goals within the peer mentoring relationship. As shared in Chapter 3, motivation is a key component to ensuring that students continue their work toward meeting their desired outcomes and is largely reliant on students' mindsets. Likewise, building trust within the relationship is essential (APA, 2012). During the relationship working phase, it will be critical for you to model to students how to develop and sustain a growth mindset. Your specific use of words—especially praise—needs to be thoughtful and constructed through the growth mindset lens. Remember the importance of praising effort—that is, hard work, persistence, refusing to give up—rather than praising innate talent or ability. While it may not always seem this way, your students are often looking to you for validation. Validate their effort so that they learn the valuable skill of adopting a growth mindset. For instance, rather than saying to a student, "You are so talented!" you could reframe the sentiment by saying, "I am impressed by how dedicated to the task you were. You persisted even when things were challenging!"

It's important to remember, though, that people (your students in particular) typically know when they have truly done a good job or not—whether they have actually excelled at a task. Thus, be mindful to use praise appropriately, providing constructive feedback and encouraging a growth mindset. Especially among students who have had difficult past experiences, providing unwarranted or even too much positive feedback or praise can be detrimental to students' sense of motivation, sense of belonging, and trust (Phillips et al., 2020). Phillips and colleagues (2020) share that adults nowadays tend to be effusive in their praise of students, mostly so that students can feel better and so that we all can avoid emotional discomfort. But doing so instead breeds anxiety, fear of failure, and works against the development of a growth mindset. I would also argue that it also makes those of us offering the effusive praise less trustworthy, which in turn damages the mentoring relationship. Instead, we want to assist our students in developing resilience and an internal motivation to pursue knowledge and skills that will enable our students' success long-term. Refer back to Chapter 3 for more information on the growth mindset.

In addition to providing support in developing a growth mindset, you will also need to monitor students' progress toward meeting their goals within the peer mentoring relationship. There are many ways that this can be accomplished. For instance, regular one-on-one check-ins can be conducted so that mentors and mentees can share with you privately how things are going, challenges that they are encountering, and requests for support. If you require mentors to complete mentoring meeting notes, reviewing those notes could be helpful in assessing momentum. A simple survey sent to students via email or text could also suffice. Another suggestion that students shared with me in one of the most recent mentoring cohorts that I facilitated was having regular group meet-ups, which could serve as a check-in with faculty facilitators as well as an opportunity for various mentor and mentee groups to check in with each other for encouragement and sharing of experiences and resources. The questions that you ask, the feedback that you provide, and the words that you use are all very important. In Chapter 5, I'll discuss more fully how you can ensure that your communication is clear, effective, and meaningful.

Finally, during the relationship working phase, you will need to foster students' ability to solve problems. If you recall, I shared earlier in the chapter that faculty facilitators are not responsible for giving students all of the answers, nor are mentors responsible for giving mentees all of the answers. Rather, a critical competency that you will be working toward teaching your students through your facilitation of the peer mentoring relationship is how to identify problems, seek creative solutions, and move toward action in solving identified problems. This is where it is very important for you to remember the concept of "guide on the side." Just as you would when teaching new information and concepts to the students in your classes, you will need to scaffold students' learning in identifying problems and challenges, brainstorming potential solutions, determining which solutions are most feasible and most likely to facilitate the desired outcome, determining steps to take action, and facilitating students taking the first steps toward solving the problem or challenge. I am confident that the majority of faculty reading this book want to see the next generation grow into competent critical thinkers and problem-solvers. It is during the relationship working phase that you will be able to assist with this! In Chapter 6, I'll talk more about problem-solving and mitigating challenges.

Vision: Encouraging Assessment through Self-Reflection

A critical part of the relationship working phase is regular evaluation of the peer mentoring relationship (Rockinson-Szapkiw et al., 2020). You will want to monitor and evaluate progress toward meeting goals, but it will also be essential for you to facilitate mentors' and mentees' self-evaluation as a form of monitoring and construction of *vision* as well.

One of the most effective ways that I have seen faculty do this is through the use of reflective journals. When faculty have mentors and mentees reflect on their journey through the peer mentoring relationship in a journal—whether paper, digital, audio, or other form—students are better able to monitor progress, reflect on areas of strength and areas in need of improvement, identify and work through challenges, and determine effective "next steps." I *do not*, however, recommend that you as the faculty facilitator read, watch, or listen to students' reflections. The reflections should be confidential and only read, watched, or listened to by the faculty facilitator if the mentor or mentee asks you to do so. Otherwise, the journal should be mentors' and mentees' own personal way of reflecting on how the peer mentoring journey is progressing as well as their own personal progress, feelings, and emotions. However, you can reiterate the importance of this self-reflective tool and encourage students to engage in journaling. This is also one way that you can support students in engaging in mutual accountability (Zachary & Fain, 2022) and continued progress through both setting and meeting goals. (For additional resources on journaling, see Chapter 7—specifically Zachary & Fain's *The Mentor's Guide*.)

Supporting the End of a Relationship (Phase 4)

The final phase of the peer mentoring relationship is the relationship termination phase. As shared previously, there are many reasons why a peer mentoring relationship might come to an end. The ending of a peer mentoring relationship is not inherently bad! Sometimes students may have met their goals and may now be ready to move on to the next phase of their journey separate from peer mentoring. Mentees, especially, may have accomplished their goals and be ready to move into the role of mentor for others—this was noted repeatedly by mentees within my own peer mentoring work. The drive for mentees to give back and reciprocate was enormous! Other times, students may be graduating, moving into the workforce, or leaving their degree program for various reasons. Occasionally, the mentor and mentee match was not appropriate—perhaps personalities differed or goals did not align—and the relationship needs to come to a conclusion. Regardless of the reason, all things must eventually come to an end, and ensuring a tidy wrap-up of the peer mentoring relationship is essential. In fact, the end of the relationship should be considered from the very beginning of the relationship and, thus, planned for appropriately (Rockinson-Szapkiw et al., 2020).

In my work in peer mentoring, I've found it useful to follow the three R's presented in *Navigating the Peer Mentoring Relationship* (Rockinson-Szapkiw et al., 2020): review, rejoice, and reflect. As the faculty facilitator, it will be your responsibility to assist students in reviewing and reflecting on the peer mentoring experience—much of which you have likely already completed through the relationship working phase. Here, you will want to support mentors and mentees in determining what goals were achieved, whether what they achieved aligned with their expectations, what they learned, and what they still hope to accomplish. This is also the time to celebrate and rejoice in all that has been accomplished—the respective contributions that both mentor and mentee brought to the relationship (no matter how large or how small), the learning and growth that has occurred, and the lessons learned that will carry forward into the future. Likewise, this is the time to express appreciation for the effort that all have put into the peer mentoring relationship—without judgment.

Finally, this is when you will further assist students in determining "next steps." That is, what will the mentor and mentee do individually to ensure momentum and progress toward remaining goals? What plan can be put in place to ensure success toward meeting these goals? This is also the time to determine whether the mentor, mentee, and faculty facilitator will simply part ways or whether continued communication is appropriate and allowable. This will be up to each respective individual and will be largely determined by needs, availability, and boundaries. However, keep in mind that when a peer mentoring relationship comes to an end, a new type of professional relationship can potentially take its place. In my experience, some of my closest professional relationships have been born in some way, shape, or form through a mentoring relationship. This also applies to the relationship between you as a faculty facilitator and your students. We'll discuss in greater detail shifts in relationship role in Chapter 6.

Summary

In this chapter, we discussed in detail the roles of the peer mentor, the peer mentee, and the faculty facilitator. We discussed methods for ensuring fit and matching peer mentors and peer mentees. The necessity of peer mentoring training for both peer mentor and peer mentee was shared as well as multiple resources, including the eSTEM peer mentoring training that I have co-developed and worked most closely with. Your expectations as a role model were explored. And your role in supporting the four phases of the peer mentoring relationship was detailed: (1) relationship building, (2) relationship negotiation and expectation setting, (3) relationship working, and (4) relationship termination. Finally, we once again touched on the importance of self-reflection, both on your part and the part of the peer mentor and peer mentee. As we wrap up this chapter and move forward in our learning, take a few moments to consider the questions and prompts presented in the Reflect and Act section.

Reflect and Act

1 Review the definition that I have provided for mentoring and, specifically, peer mentoring. Consider the myriad ways in which you believe an effective peer mentoring relationship should be reciprocal. What are the benefits overall for approaching relationships as reciprocal? What challenges, if any, might be posed?
2 Consider how your students may look to you as a role model. In what ways might they look to you for guidance?
3 What characteristics do you believe make a good mentor?
4 What characteristics do you believe make a good mentee?
5 What characteristics do you believe make a good faculty facilitator? Which of these characteristics do you show? Which of these characteristics do you need to more fully develop? How might you attend to this?
6 Review the various roles of the peer mentor, peer mentee, and faculty facilitator. Which of these roles do you think will be most challenging within the context in which you will be facilitating the peer mentoring relationship? How might you work toward mitigating these challenges?
7 Consider the personalities, strengths, weaknesses, needs, goals, areas of disciplinary interest, and personal characteristics of your students. Are there particular areas in which a match or fit is clear? Are there areas in which a match or fit is unclear or challenging? How might you work with your students to determine the most appropriate match?
8 What are the benefits of match and fit? What are the benefits of diversity within the peer mentoring relationship?
9 Take some time to explore the peer mentoring training resources presented in this chapter. How might you utilize the training to support your students as they enter into the peer mentoring relationship?
10 Consider the four phases of peer mentoring. With which of these phases are you most comfortable? With which of these phases do you feel least comfortable? Make an action plan to gain the knowledge, resources, and level of comfort that you will need to be able to appropriately facilitate each of the four phases.

11 Reflect on the boundaries that you will need to set, communicate, and hold—both personally and professionally—as you enter your role as a faculty facilitator.
12 Brainstorm a list of networks, professional organizations, conferences, workshops, listservs, and research opportunities that you have access to and might be able to share with your students to support their professional growth.
13 Consider how you will monitor the overall health and effectiveness of the peer mentoring relationship over time. Create an action plan using the three R's: review, rejoice, and reflect.

References

Akinla, O., Hagan, P., & Atiomo, W. (2018). A systematic review of the literature describing outcomes of near-peer mentoring programs for first year medical students. *BMC Medical Education, 18*(98). https://doi.org/10.1186/s12909-018-1195-1

APA. (2012). *Introduction to mentoring: A guide for mentors and mentees.* https://www.apa.org/education-career/grad/mentoring

Daloz, L. (1999). *Mentor: Guiding the journey of adult learners.* Jossey-Bass.

eSTEM Mentoring. (2025). *eSTEM Peer Mentoring.* https://estemmentoring.com/

Jones, V. O., & Wendt, J. L. (2025). Encouraging confidence: The impact of an online peer mentoring program on women peer mentees in STEM at two HBCUs. *Trends in Higher Education, 4*(3). https://doi.org/10.3390/higheredu4010003

McGee, E. O. (2016). Devalued Black and Latino racial identities: A by-product of STEM college culture? *American Educational Research Journal, 53*(6), 1626–1662. https://doi.org/10.3102/0002831216676572.

Miller, D. (2021). *Honoring identities.* Rowman and Littlefield.

Mondisa, J. (2018). Examining the mentoring approaches of African-American mentors. *Journal of African American Studies, 22,* 293–308.

Pfund, C., Byars-Winston, A., Branchaw, J., Hurtado, S., & Eagan, K. (2016). Defining attributes and metrics of effective research mentoring relationships. *AIDS and Behavior, 20,* 238–248.

Phillips, S., Melim, D., & Hughes, D. A. (2020). *Belonging: A relationship-based approach to trauma-informed education.* Rowman & Littlefield.

Rockinson-Szapkiw, A., Wendt, J. L., & Wade-Jaimes, K. (2020). *Navigating the peer mentoring relationship: A handbook for women and other underrepresented populations in STEM.* Kendall Hunt Publishing Company.

Rolfe, A. (2021). *Mentoring mindset, skills and tools.* Mentoring Works.

Saffie-Robertson, M. C. (2020). It's not you, it's me: An exploration of mentoring experiences for women in STEM. *Sex Roles, 83,* 566–579.

Taylor, K.-Y. (2017). *How we get free: Black feminism and the Combahee River Collective.* Haymarket Books.

Wendt, J. L., & Jones, V. O. (2024a). Peer mentors' experiences in an online STEM peer mentoring program: "Beacons of light." *International Journal of Mentoring and Coaching in Education, 13*(3). https://doi.org/10.1108/IJMCE-03-2023-0033

Wendt, J. L., & Jones, V. O. (2024b). Supporting BIPOC males in STEM: Insights from a case study on online peer mentoring. *Journal of Research in STEM Education, 10*(1–2), 89–113. https://doi.org/10.51355/j-stem.2024.145

Wendt, J. L., Jones, V. O., & Bruewer, A. (2021). Understanding the peer mentoring experiences of STEM mentees at two HBCUs. *The Chronicle of Mentoring and Coaching, 5*(14), 487–491.

Zachary, L. J., & Fain, L. Z. (2022). *The mentor's guide: Facilitating effective learning relationships.* Jossey-Bass.

5
LEADING BY EXAMPLE

Introduction

A key skill to being an effective leader and, in the case of peer mentoring relationships, an effective faculty facilitator, is being able to lead by example. We've all heard the old adage, "Do as I say, not as I do." But in reality, effective faculty facilitators should take the opposite approach and quite literally practice the words that they preach. That is, you must actually do the things that you tell your student mentors and mentees to do—model the skills and competencies that you want for them to learn! After all, modeling is the primary way through which human beings learn. Let's examine learning theory to better understand your role as a leader.

Learning through Social Interactions

If we take a minute to consider the process of human development and the key concepts of learning theory, Sociocultural Theory (Vygotsky, 1978b, 1986) is grounded in the fact that human beings are social creatures (see Chapter 1). From our first moments of life, we begin a lifelong journey of learning through observing and interacting with other people. It is through our observations and interactions with others that we internalize new information and begin developing independent thought. Sociocultural Theory highlights the importance of learning through our interactions with others—specifically with a "skilled helper" (Levine & Munsch, 2022, p. 42). A skilled helper during the phases of child development might be a parent or teacher. In the context of peer mentoring during post-secondary education, the skilled helper might be the peer mentor (which would typically be a student with slightly more advanced knowledge in one or more areas) or might be the faculty facilitator. Regardless, we look to others—both in their words and actions—to gather new information, to learn new skills, to learn new competencies, and to learn how to better navigate the world. And the skilled helper is the individual that helps us accomplish our greatest gains in learning at the beginning of our journey.

A key concept within Sociocultural Theory is that of the *zone of proximal development* (Vygotsky, 1978b). The zone of proximal development is defined as "the distance between the actual developmental level as determined by independent problem solving and the level of potential development as determined through problem solving under adult guidance or in collaboration with more capable peers" (Vygotsky, 1978a, p. 86). Essentially, the zone of proximal development within the context of education and learning is important because it is the space within which learning and growth actually occur. It is that space that exists in the delicate balance of what a person knows, the challenge that an effective educator provides, and what a person *can* know or do. It is that space known as *potential*. The challenge that an effective educator provides must draw on what the student already knows and can do and, through scaffolding, pushes the student just outside of their comfort zone so that they can begin the construction of new knowledge, skills, and competencies. See the Venn diagram in Figure 5.1 as a visualization of the zone of proximal development.

As a faculty facilitator, it is your job to help challenge students—just as I shared in Chapter 4—throughout the peer mentoring relationship by identifying and pushing students into that zone of proximal development. More

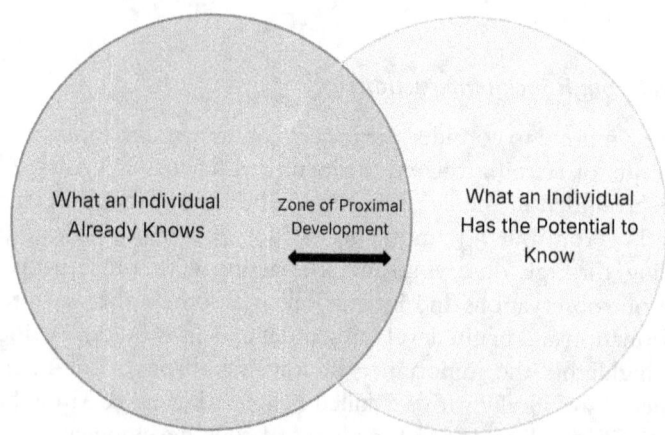

FIGURE 5.1 A visualization of the zone of proximal development.

specifically, the challenges that you provide, which you can also think of as opportunities for growth, will be part of the relationship working phase (see Chapter 4 for a description of each of the phases of the peer mentoring relationship). These opportunities for growth will come through you encouraging and pointing students toward learning new discipline-related content knowledge (academic growth), through you sharing professional development opportunities—internships, conferences, access to professional organizations, access to professional networks (social growth), and through you encouraging the adoption of a growth mindset (psychosocial growth). Further, by serving as a skilled helper, you will also model what it looks like to be a successful, competent professional in your field. You will model the habits and skills that a professional within your specific discipline area adopts. You will serve as a role model for your students—a role model through competent leadership. Let's now look at what it means to be a competent leader.

Being a Competent Leader

Competent leadership is a cornerstone to facilitating effective and meaningful peer mentoring relationships. As the faculty facilitator, it will be your responsibility to lead and guide your students in a way that assists them in collaboratively working toward meeting their goals—both individually and together within the peer mentoring team. A key component of being a competent leader is being able to assess and reflect on strengths and areas that could be improved. Rolfe shares that "for a leader, understanding your own strengths and those of the people you lead is essential" (2021, p. 98). Thus, it is essential to not only take time to reflect on your own strengths but also to focus on enhancing and building upon those strengths. Keep in mind that, while weaknesses (more preferably referred to as "areas in need of improvement") are important, the focus should be more of a strengths-based focus than a deficit-minded approach (Wells, 2024)—both for yourself as well as for your student mentors and mentees. (A deficit-minded approach tends to focus on what you can't do well at the moment or, in other words, where you lack.) We all have strengths, and we all have weaknesses (as well as areas in which we can learn and improve), but when we know what our strengths are we can more intentionally capitalize on those strengths in order to work in collaboration with others.

The same concept applies to our students. There is no benefit within peer mentoring relationships in adopting a "weakness-focused, punitive approach" (Rolfe, 2021, p. 97). A strengths-based approach—one that encourages confidence and growth–is derived from developmental psychology, which adopts the view that each individual has unique natural strengths and talents that inherently enable them to reach and maximize their fullest potential (Rolfe, 2021). This also aligns with the concept of growth mindset in that our abilities and talents are malleable and largely dependent on our own beliefs of self (Dweck, 2017; Dweck & Yeager, 2019). Importantly, it further aligns with the reciprocal nature of the peer mentoring relationship, which recognizes the unique experiences and attributes that each individual has to offer.

So, how do you show up as a leader? Do you model sound practices? Do you challenge barriers? Do you push your students gently outside of their comfort zone in thoughtful and intentional ways to promote growth? What words do you say and what messages do you convey to students? These questions dovetail nicely with our methods of communication, so let's start there.

Effective Communication

Communication is considered "the essential building block for facilitating all learning relationships" (Zachary & Fain, 2022, p. 47). This is especially applicable to your ability to facilitate effective peer mentoring relationships. You must be able to clearly communicate your expectations of students and your suggestions for action in a way that encourages students to continue forward momentum toward meeting goals while simultaneously demonstrating empathy, respect, and cultural responsiveness. A key component of effective communication is not just active (intentional) listening, but empathic listening.

Empathic Listening

Most of us when asked would readily confirm that we are good listeners, especially as faculty who work with students on a daily basis. But are we really? Do we truly listen—truly hear what others are telling us? Or are we processing the minimum amount of information before formulating a response in our mind, or, worse yet, interrupting with our all-knowing wisdom? I would argue that faculty—especially those in academia—would benefit from practicing the art of empathic listening more intentionally.

Empathic listening is the skill of listening actively and for understanding (Covey, 2012). Through the process of empathic listening, we silence our own thoughts and perspective, truly listen to what others are sharing with us, and maintain our focus on the other person. We do this through inserting very minimal responses, taking appropriate time to reflect on what is said, and asking questions to ensure deep understanding. Rolfe (2021) shares in her book *Mentoring: Mindset, Skills and Tools* a fabulous

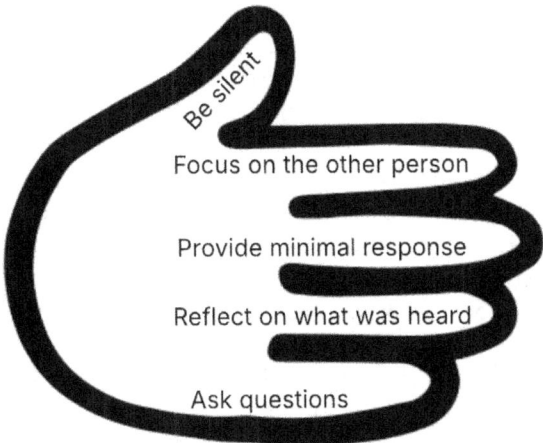

FIGURE 5.2 A visualization of the five steps of empathic listening.

Source: From *Hand, gesture icon*, by khushmeen icons, Iconfinder. (https://www.iconfinder.com/icons/9165553/hand_gesture_icon). CC BY 4.0.

tool for remembering the strategy of empathic listening. She suggests that you visualize the strategy of empathic listening as five steps, conveniently represented by the five fingers on your hand. The first step—represented by your thumb—is being silent. The second step—represented by your forefinger—is focusing on the other person. The third step—your middle finger—is providing minimal response. The fourth step—your ring finger—is to reflect on what you have heard. And the final step—your pinky finger—is asking questions for deep understanding. This can be visualized in Figure 5.2. As you practice the skill of empathic listening, use your hand as a visual reminder of the steps that encompass effective empathic listening.

I also have found it helpful to understand the various levels of listening (also keenly pointed out by Rolfe (2021) in her book) as presented by Covey (2012). Covey, in his book *7 Habits of Highly Effective People*, organizes listening into five distinct and progressive levels. At Level 1, the least effective level of listening, we ignore what the other person is saying. At Level 2, we pretend that we are actively listening to the other person. Perhaps we nod or mutter an "um hmm." Regardless, we are not truly engaged. At Level 3, we listen selectively. That is, we only pay attention to a portion of what is being said. This is the stage where we often become frustrated or impatient, feeling the urge to interrupt and move the conversation along. At Level 4, we attentively listen to the other person. We give them our attention, but our mind is somewhere else—on judging and assessing what the other person is telling us based on what we value, what we need, and what we prioritize. We might feel the need to advise and provide our input, focusing on our own perspective on the topic. At Level 5, though, we practice empathic listening. This is when we truly give our focused attention to the other person, hearing and deeply understanding what they are communicating, and doing so through their perspective—not our own. A Level 5 is our goal as a faculty facilitator engaging in empathic listening. Understand, though, that this is a learned skill (Covey, 2012)—one that requires continual practice and self-monitoring. It is also a skill that very few of us find to be easy or natural. Figure 5.3 provides a visualization of the five levels of listening.

Through empathic listening, you will also be able to begin building trust within the peer mentoring relationship. Mentoring is inherently a delicate process, a process that introduces a certain level of vulnerability as it requires transparency, honesty, and openness (APA, 2012). However, if trust is not established, the mentoring relationship—along with all of the desired outcomes and goals—will be stunted. Let's now explore how communication ties to trust.

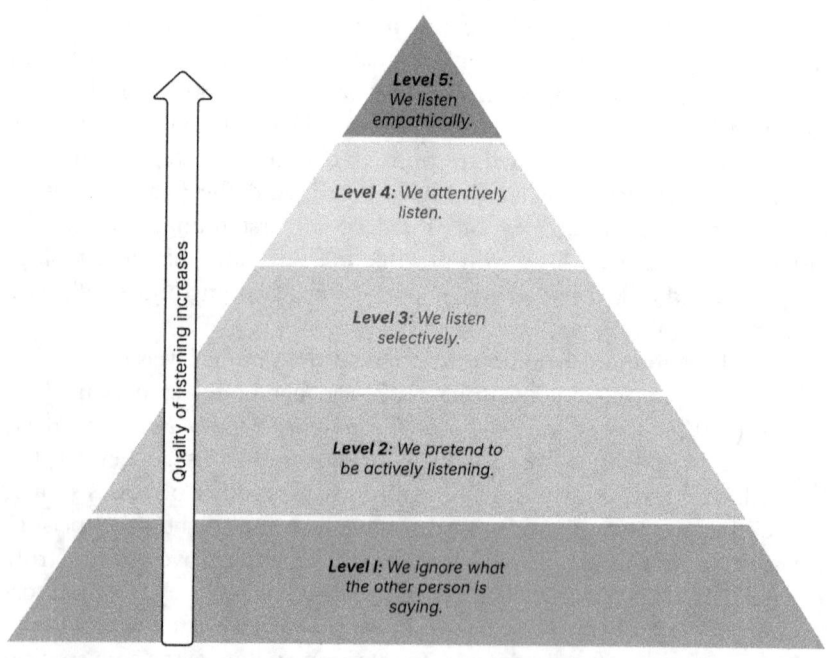

FIGURE 5.3 A visualization of the five levels of listening.

Building Trust

Trust can be defined as "the extent to which a person is confident in, and willing to act based on, the words, actions, and decisions of another" (McAllister, 1995, p. 25). Trust takes time and is rarely given immediately. If you already have a relationship with your students prior to the peer mentoring relationship, then you may have a head start on building trust. Otherwise, be prepared for trust building to take time and to be a gradual process—one that is only developed with intentionality.

Unconditional Positive Regard

One method for developing the skills necessary for building trust is through the practicing of unconditional positive regard, also known as UPR (APA, 2012; Rogers, 1957, 1959). *Unconditional* means that you as the faculty facilitator will accept your students as mentors and mentees unconditionally—that is, with no conditions and no judgment. *Positive* means that you as the faculty facilitator will foster a feeling of value and "warm acceptance" (Rockinson-Szapkiw et al., 2020, p. 38). *Regard* means that you as the faculty facilitator will respect your students as mentors and mentees with individual autonomy. That is, you will respect them as their own individual human beings. Thus, UPR means that acceptance and respect are afforded and that judgment and evaluation are withheld. By practicing UPR in your communications and actions with your students, you can help to develop a relationship built on trust. Empathic listening is one way that you can practice UPR.

Empathy

Another way to develop trust is by practicing empathy. Empathy is a method for communicating understanding, letting your students know that they are valued, seen, and heard. Empathy can be defined as "the ability to 'see' and 'feel' each other. It means to 'feel into' or 'feel with'. It is communicating 'I get it' or 'I understand'" (Ivey et al., 2014 as cited in Rockinson-Szapkiw et al., 2020, p. 39). Empathy is when you "fully, deeply, understand [a] person, emotionally as well as intellectually" (Covey, 2012, p. 396). Empathy can be communicated through non-verbal expressions, such as nodding in understanding, or through verbal expressions such as validating what a student shares with you. Validation could be you sharing "I understand that you feel ..." or "That sounds like it is very challenging for you." It, however, does not mean taking your students' problems as your own, nor does it mean solving the issue for your students. Rather, coupled with empathic listening (which holds empathy at its core!) (Covey, 2012), it is a way to communicate value and understanding while simultaneously guiding a student to engage in critical thinking and problem-solving.

Emotional Intelligence

Hand-in-hand with empathy, another competency that you will need to master and model as a faculty facilitator is emotional intelligence. Emotional intelligence is "the ability to recognize and understand our own emotions (self-awareness) and the emotions of others (social awareness) and then use this ability to guide our behavior (self-management) and manage our relationships (relationship management)" (Zachary & Fain, 2022, pp. 5–6). Each of us possesses a particular level of emotional intelligence, but that level is dynamic and can improve with practice and time. For instance, becoming more in tune with your own emotions and moods can lead to enhanced self-awareness, as can developing an understanding of how your emotions and moods might impact other people (Zachary & Fain, 2022). Your ability to refrain from judgment, for instance, can improve your levels of self-management. Likewise, developing flexibility in changing environments (like those we often find in higher education) can enhance your self-management. Social awareness can be enhanced by being aware of and taking into consideration others' experiences, needs, perspectives, and emotions. Relationship management can be enhanced by practicing appropriate responses to others' emotional reactions. Importantly, relationship management can also be improved by engaging in cultural sensitivity through culturally responsive practices (refer back to Chapter 3 for a more robust discussion of cultural responsiveness). This leads me next to a discussion on the importance of understanding context.

Understanding Context

A key competency of an effective leader and, thus, an effective faculty facilitator is the ability to recognize and understand context. Context will no doubt impact everything that you do as a faculty facilitator. Context can be described as "the circumstances, conditions, and contributing forces that affect how we connect, interact with, and learn from one another" (Zachary & Fain, 2022, p. 33). It is complex, multi-layered, and will change continuously depending on how, with whom, why, where, and when you are interacting with your student mentors and mentees. It may even change within the course of one particular mentoring relationship, as mentoring relationships (and, thus, contexts) are flexible (APA, 2012). In Zachary and

Fain's (2022) book *The Mentor's Guide: Facilitating Effective Learning Relationships*, two very detailed chapters on context within the mentoring relationship are presented—I strongly recommend you consult this resource if you find that a deeper dive into context would be beneficial to you. Here, though, I'll try to keep things as simple as possible, starting with context based on the relationship structure.

Relationship Structure

The first thing to consider regarding context is the particular structure of the mentoring relationship that you will be facilitating. This book, of course, focuses on your facilitation of the peer mentoring relationship, but there are myriad other mentoring relationship structures within which you may find yourself (see Chapter 1). You need to determine the nuances surrounding the relationship structure in which you are involved and are facilitating. For instance, is the peer mentoring relationship that you are facilitating a formal relationship or an informal relationship? How much structure is embedded within the relationship? What is the overall purpose of the relationship? What are the goals that the peer mentor and peer mentee have set for the relationship? And what is your role in facilitating your students in meeting those goals? The structure of the relationship will be different for each relationship that you are facilitating, will be dependent on the particular individuals involved and their respective needs, and may fluctuate as goals and needs are modified. You can think of the context of the mentoring relationship as being a bird's eye view of the relationship—the overall big picture (Zachary & Fain, 2022).

Personal Characteristics

The next part of understanding context has to do with recognizing the individuals involved in the peer mentoring relationship—that is, the peer mentor, the peer mentee, and the faculty facilitator (you!)—as autonomous human beings. Zachary and Fain (2022) refer to this as "the context of difference" (p. 36). Each of us is different. We come to the table with different experiences, cultures, identities, religions, worldviews, gender identities, and sexual orientations. Within the mentoring relationship, we might even hail from different generations, introducing a generational context to the mix. Each of these characteristics impacts who we are and how we navigate the world around us. Personal characteristics "shape our thinking, our conversation, our relationships, and our behaviors" (Zachary & Fain, 2022, p. 37). They form the lens through which we view the world and our interactions with others. Thus, it is imperative that you take time to interrogate that lens and to reflect on how your own personal characteristics may influence how you view the world and how you interact with your students. Likewise, it is important for you to take the time to get to know your students so that you can understand the lens through which they see the world. This ties directly back to our previous discussion on cultural responsiveness (see Chapter 3). I would also encourage you to check out the resources in Chapter 7 to further support your interrogation of context.

Power Dynamics

Another important aspect of context that must be considered is that of power dynamics. Power can be defined as "the potential influence that one has over another person or group" (Kovach, 2020, p. 102). It is often the result of a hierarchical relationship—such as that which may occur naturally within a peer mentoring relationship. If you recall, in Chapter 1, I shared the following definition of mentoring with you: "a process in which an experienced individual (a mentor) provides emotional and psychosocial support (e.g., listening, empathizing, offering advice, providing affirmation or an objective perspective), and helps to educate, guide, and counsel a less experienced person" (Mondisa & Adams, 2022, p. 339). By examining the definition of mentoring that I have adopted within this book, it is obvious that there is likely some power difference between the peer mentor and peer mentee just by sheer nature of experience and knowledge garnered from such experience. This power difference is known as *power differential*.

Just as a power differential is likely to exist between peer mentor and peer mentee, there will also inevitably be a power differential between you as the faculty facilitator and your students within the peer mentoring relationship. You as the faculty facilitator will, of course, hold a higher level of knowledge due to your experience and role as an expert within your field. Let me be clear, though. This does not mean that you have nothing to learn from your students! We *all* have something to learn. *All the time.* This will never change. It is part of the human experience. But inevitably, you as the faculty facilitator will occupy a different level of the hierarchical structure within the peer mentoring relationship simply by nature of being the faculty facilitator (see Figure 5.4). It is imperative that you are aware of this.

By occupying a higher level of the hierarchical structure, you as the faculty facilitator will hold more power within the peer mentoring relationship. This can be a good thing that can facilitate the relationship, but it could also be a bad thing that could stunt or even damage the relationship. The power differential that exists must be recognized and managed accordingly in a way that serves to support your students as they navigate the peer mentoring relationship. Let me explain.

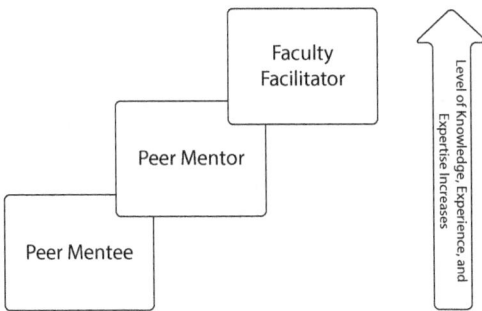

FIGURE 5.4   A visualization of the inevitable hierarchical structure of the peer mentoring relationship.

A healthy peer mentoring relationship is one that is developmental. That is, it assists those within the relationship with gaining the acceptance, affirmation, and opportunities to be challenged that will result in development or, in more simple terms, learning and growth (Beech & Brockbank, 1999). When such opportunities exist, a developmental alliance (Clarkson & Shaw, 1992) between those within the peer mentoring relationship is established. As Beech and Brockbank purport, "active learning and development are likely to emerge from developmental relationships, where they occur, as such relationships foster autonomy and independence rather than passivity and dependence" (1999, p. 8). When a power differential exists and is not actively checked, it can result in the developmental alliance being unequal, damaged, or stunted. That is, if you as the faculty facilitator are not cognizant of the power that you hold in the relationship in comparison to the power that your student peer mentor and peer mentees hold in the relationship, you run the risk of limiting the extent to which the peer mentor and peer mentees can engage in active learning, leading to superficial learning and, thus, less growth (Beech & Brockbank, 1999; Freire, 1972). Beech and Brockbank (1999) refer to this as the *hierarchical boundary*. The hierarchical boundary can lead to challenges such as dependency and co-dependency (as well as other challenges) that will result in the peer mentoring relationship being less effective than it could potentially be.

So, how do you as the faculty facilitator appropriately manage the power differential within the peer mentoring relationship? The first thing to do is to recognize that, while you might have more experience and expertise in a particular area, you don't know everything. You haven't lived the experiences that your students have lived. You don't know their histories, what their home lives have looked like, the challenges and difficulties that they have encountered in their experiences, or the macro- and microaggressions and inequities that may have impacted them. You also don't know what knowledge they may have gained from their unique experiences. So, approaching the facilitation of the peer mentoring relationship through the lens of empathy and listening is key. While you will be serving as a transformative leader in the relationship, it is essential that you lead in ways that prevent others from feeling less-than (Ferris & Waldron, 2022). It is essential that you lead in ways that seek to bring value to your students, rather than devaluing them.

This means that you'll need to do more empathic listening than advice giving, only stepping in when absolutely necessary, and also keeping a keen eye on your cultural responsiveness. Practicing unconditional positive regard will also assist with this. Your experiences are not your students' experiences and vice versa. Be mindful of the lens through which you are viewing your students, their needs, and their goals. Rather than prescribing a path with specific steps for your students to follow, ask questions, guide accordingly, and provide the necessary directions to appropriate resources. Then, let your students do the work! Let them make the decisions. Let them make mistakes. And let them learn.

As I tell my young daughter almost every single day, mistakes are okay. No one learns without making mistakes. They are an essential part of the learning process, which leads not only to learning and growth but also to resilience and strength in the face of challenges and difficult situations. Mistakes help us develop grit and, hopefully, a growth mindset. Ultimately, that's what we are trying to assist our students within the peer mentoring relationship in gaining. Once they have that, the rest of the skills and competencies necessary to meet their goals within their respective fields will come more easily.

Sometimes, though, this is easier said than done. Thus, it will be essential for you to set and maintain personal and professional boundaries. This will enable you to support your students effectively while simultaneously remaining mindful of your own well-being.

Boundaries

A key component of ensuring that you are effective as a faculty facilitator is setting and holding boundaries. Boundaries are what psychologist Nicole LePera describes as a self-healing practice that allows you to focus on protecting your levels of stress, peace, and energy (2021). Boundaries are clear, concise limits of what you are and are not willing to provide, support, and/or engage with. Establishing and maintaining boundaries is absolutely essential in ensuring that you are able to support the learning and development of your students within the peer mentoring relationship while simultaneously supporting your own professional balance, peace, and self-care. I discussed previously the challenges with faculty having limited time, limited support, and, thus, being more at risk of burnout. And one of the purposes of this book is to help you in gaining the knowledge and skills to effectively support peer mentoring relationships without increasing the burden on you and, thus, the likelihood of burnout.

Boundaries are rooted in the idea that we can only control ourselves—we can't control others (although many of us try to, which is a conversation for a completely different book!). Boundaries provide the limit, the hard line for what we are willing and able to give and what we are willing and able to engage with. In the context of the peer mentoring relationship, it will be essential for you to set boundaries that support the growth and development of your students while simultaneously protecting your own time, energy, and peace. This might look like setting specific office hours, setting specific ways in which students can communicate with you (for instance, email, phone call, or text message), as well as the amount of time and effort you are able to expend on seeking out and providing resources for students. You are only one person, and you have a lot on your plate already. You'll need to decide where to draw the line so that you are appropriately supporting your students while also protecting work-life balance. As Wolfe shares, "self-preservation is a strength, not a liability" (2022, p. 31). You must take care of yourself first if you expect to be able to care for anyone else.

This extends to my previous discussion about letting students be the driver for decisions and letting them make mistakes. As Beaudoin and Maki share in their book *Mindfulness in a Busy World: Lowering Barriers for Adults and Youth to Cultivate Focus, Emotional Peace, and Gratefulness* (2021), "we are blessed with little power over the outside and immense power over our inside" (p. 100). That is, we can't control what happens around us just as we can't control the actions and reactions of those around us. But we can control our own actions and reactions. It is up to us to determine how and to what extent we want to act or react. It is also up to us to allow other people—in this case, our student peer mentors and peer mentees—to make their own decisions.

This is reiterated by Mel Robbins in her best-selling book *The Let Them Theory: A Life-Changing Tool That Millions of People Can't Stop Talking About* (2024). In her book, Robbins reminds us that as independent human beings, other adults are allowed to have their own thoughts, opinions, and feelings and are allowed to make their own decisions about how to act or react to a given situation. It is our responsibility to manage our own selves—not others. Thus, let your students figure things out. By all means, give them the tools and competencies that they need to do so. But ultimately, it is up to them to act on those tools and use those competencies to make their own decisions and chart their own course to success. You are simply there as a faculty facilitator to guide them and to facilitate the overall process. Of course, we want to see our students succeed! But it is ultimately up to them to do what is necessary to achieve success. It is up to us to give them the time, space, resources, and opportunities that support whatever success looks like to them.

Remember—you are the "guide on the side" as the faculty facilitator. You are not your students' parent, counselor, or therapist. There is often a fine line between facilitating a student to be a proactive and competent future expert without engaging in behaviors that instead promote co-dependency. Hold the line for where your boundaries stand and, if it is helpful, review the roles and responsibilities of the faculty facilitator as presented in Chapter 4 to recalibrate as needed. With this in mind, let's now talk about how to effectively manage conflict.

Managing Conflict

As a faculty facilitator, it will be inevitable that you will be placed in the position of assisting your mentors and mentees in managing conflict at some point in time. Conflict occurs in all relationships, regardless of mentoring type or context (Zachary & Fain, 2022). Conflict can be defined as "an expressed struggle between at least two interdependent parties who perceive incompatible goals, scarce resources, and interference from others in achieving their goals" (Wilmot & Hocker, 2001, p. 11).

The most important thing to remember when encountering conflict is that managing the conflict does not mean that differences will be eliminated. Rather, it means that an honest, open sharing of opposing experiences, views, and needs will occur and some sort of negotiation or compromise will occur, which hopefully involves some level of understanding and a high level of respect and compassion. "It is about inviting dialogue to understand varying points of view" (Zachary & Fain, 2022, p. 111). When conflict occurs, it is not your responsibility to solve the problem for your students. Rather, it is your responsibility to support them in engaging in effective dialogue and finding appropriate solutions.

When I am not researching and leading peer mentoring programs, I am teaching in a teacher preparation program. Several of the many courses that I routinely teach are classroom management courses. While conflict at the early childhood through secondary levels will likely look a bit different than conflict with students in post-secondary programs, the basic ideas for managing conflict are the same. In preK-12 education, we rely on restorative discipline, which is rooted in restorative justice practices (Amstutz & Mullet, 2005; Jones & Jones, 2021). *Restorative discipline* is defined as "a disposition, a mindset, and an approach to discipline that builds upon the foundational idea that schools are places where students are expected to make errors and learn from them. These errors may be both in the learning of content and in learning how to be a good member of the school community" (Milner et al., 2019, p. 133). Within the peer mentoring relationship, the mentoring relationship is essentially a community. While you won't be imparting any sort of "discipline" within your role as a faculty facilitator, you will be assisting in managing the peer mentoring relationship—the community—through conflict in ways that promote reconciliation and growth. Likewise, the practices of restorative discipline are rooted in the desire to impart "justice and equity" (Milner et al., 2019, p. 133)—age is irrelevant. The practices of restorative discipline promote communication, dialogue, sharing of impact, and collaboration to find and move toward solutions. The entire idea is to remove punishment or discipline and work together to find ways to learn and grow—and this applies no matter the age or level of education. This is exactly why I believe that restorative practices are exactly what is needed to manage conflict within the peer mentoring relationship. So, let's explore.

Restorative Practices

Within the implementation of restorative practices, there are five main goals: (1) to build positive relationships; (2) to reduce and prevent behavior that causes harm; (3) to resolve the conflict through holding people accountable; (4) to repair harm that has been caused; and (5) to attend to the needs of the community (Milner et al., 2019; Advancement Project, 2014). There are three main strategies to assist in meeting the five main goals. The strategies can be visualized as a pyramid (see Figure 5.5), as one strategy supports the others, together building a resolution (Milner et al., 2019).

The first strategy is using *affective language*. Affective language is used to share how an action has impacted or affected a person and includes honesty and openness in sharing how the person feels. The use of affective language when managing conflict should include how the person feels and was affected, what the specific action was that caused the impact, and an authentic expression that combines both feelings and action and directs the other person to the desired action (Milner et al., 2019). For instance, let's say that a mentor has repeatedly shown up late to scheduled meetings. I might use affective language to share the following: "I am very frustrated that you keep missing your meetings with me. When you are late, it throws my schedule off and causes my other appointments to run behind, which isn't fair to me or my other students." Using this language, I have expressed how I feel, what the behavior or action was that caused a conflict, how the

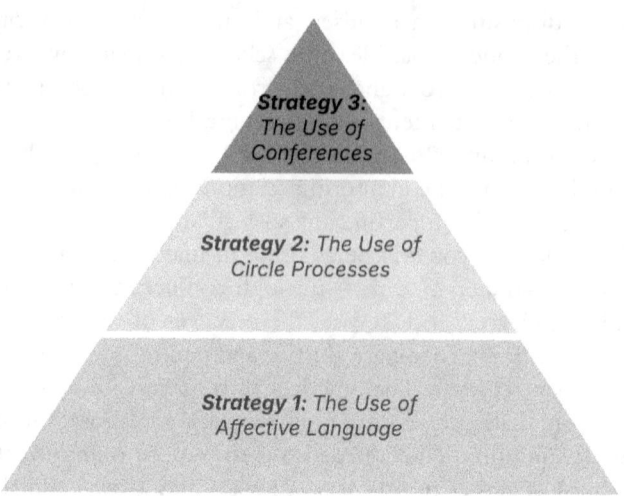

FIGURE 5.5   The three main strategies for supporting restorative practices.
Source: Adapted from Milner et al. (2019).

behavior impacts me and others, and directed the mentor toward the desired behavior—being on time. Affective language can also be used in the form of questions, asking the individual to reflect on and mindfully consider how their actions have impacted others.

The second strategy is using *circle processes*. The purpose of circle processes is to strengthen relationships and build community (Milner et al., 2019). Interestingly, circle processes have been an integral part of problem-solving in many cultures, including Indigenous cultures (Davis & Coopes, 2022; Yazzie, 1994), with much success. In a circle process, the community—in this case—those individuals involved in the peer mentoring relationship—sit in a circle so that everyone can be seen and everyone can be heard. The faculty facilitator leads the meeting by introducing the purpose of the meeting and providing a reminder of the importance of honesty, empathy, and respect. A talking piece is used to indicate whose turn it is to speak. If you are not holding the talking piece, then you simply sit and listen empathically. The faculty facilitator hands the talking piece to the first individual to speak, often posing a question to help the person begin. The talking piece is passed around the circle, giving everyone a chance to share openly. If an individual does not want to share or speak, they simply pass the talking piece along. The main questions that the faculty facilitator will want to ensure are attended to are the following: "What happened?:; What harm resulted?; What needs to happen to make things right?" (Costello et al., 2009, p. 44). When the dialogue has come to a productive, natural end, the faculty facilitator then closes the circle process. If needed, time is provided for everyone to quietly contemplate and reflect on the conversation.

A slight modification to circle processes is needed in the case of faculty facilitator involvement in a grievance or wrong. If the faculty facilitator is involved in the conflict, then a non-biased additional party should be invited to lead, facilitate, and close the circle process. It would not be appropriate for you as the faculty facilitator to lead the circle process if you are actively involved in the issue at hand above and beyond being the faculty facilitator for the peer mentoring relationship. For instance, if a mentor or mentee has a grievance with the faculty facilitator, or perhaps the faculty facilitator needs to remedy a wrong that was committed toward them by a mentor or mentee, an additional party should be brought in. This will ensure that someone removed from the situation can objectively facilitate the discussion and collaborative problem-solving process.

The final and third strategy is *conferences*. The idea behind conferences is to provide individuals the opportunity to meet face-to-face to discuss what happened, why, what the impact was, the feelings involved, and potential paths to resolving the conflict (Milner et al., 2019). Conferences are typically used only when a conflict is serious enough to have damaged or broken a relationship and involve only the impacted parties—not the entire community. Key to conferences is forging a plan or agreement for how to prevent the situation from occurring again. There are four main requirements for a conference to be effective, though: (1) participation is entirely voluntary; (2) the individual who caused the harm must have accepted at least some responsibility for the conflict; (3) the harm caused must be acknowledged by all individuals involved; and (4) an agreement is forged that outlines how the conflict will be avoided moving forward (Milner et al., 2019; Zehr, 2015).

Throughout the use of restorative practices, it is essential that not only openness and honesty be shown, but that respect for all individuals involved is demonstrated. The idea is not to point fingers or cause shame, but rather to acknowledge how actions and behaviors impact others and work toward finding ways to work collaboratively with care, concern, and kindness. After all, the peer mentoring relationship is reciprocal. Thus, respect must be reciprocal as well. With this in mind, let's now move to discussing ethics in mentoring.

Ethics in Mentoring

An important aspect of mentoring that we have not yet discussed in depth is that of ethical mentoring. Ethical mentoring can be defined "as a mentoring relationship that is intentional to guarantee the psychosocial support and well-being of both the mentor and mentee" (Villanueva et al., 2020, p. 184). The mentoring relationship is a special one—one that requires honesty and openness, which inevitably may lead to personal and perhaps sensitive information being shared. As I have shared throughout this book, it is very important to remember that your students are human beings—they have feelings, emotions, needs, desires, and dreams just like you do. They deserve empathy and kindness. And above all, they deserve your respect. Thus, ethical mentoring is absolutely imperative. While no one set of ethical guidelines for mentoring yet exists, the most commonly applied guidelines are those published by the American Psychological Association (APA, 2012, 2017).

The APA suggests that a set of General Principles (APA, 2012, 2017) be applied to mentoring relationships to guide mentors, mentees, and, in your case, faculty facilitators. There are five General Principles: (1) Beneficence and Nonmaleficence; (2) Fidelity and Responsibility; (3) Integrity; (4) Justice; and (5) Respect for People's Rights and Dignity. While you are strongly encouraged to review the APA resources cited here as well as shared in Chapter 7, I'll provide a summary.

Beneficence and Nonmaleficence

Within the context of facilitating peer mentoring relationships, Beneficence and Nonmaleficence promotes the idea of doing no harm (APA, 2012, 2017). That is, the peer mentoring relationship is reciprocal. All individuals should be engaged in a balanced give and take. This means being mindful of workload, expectations, and boundaries, for instance, ensuring that each individual in the relationship is treated with care and respect and that the relationship is one of mutualism and collaboration. This also means making sure that credit is given where credit is due. As a faculty facilitator, you will need to be cognizant that your role as faculty does not overshadow the effort expended by your student mentors or mentees. This is particularly relevant when mentors and mentees are involved in research projects that you oversee, publication opportunities that you lead, and presentation opportunities that you head. While you will inevitably be the lead on substantial projects given your advanced experience and expertise (even if not named on paper as principal investigator, first author, or lead presenter), you will need to make sure that the credit provided to your students is proportionate to the amount of time and effort they have expended. Give them a chance to shine!

Fidelity and Responsibility

This leads right into the second Guiding Principle: Fidelity and Responsibility. Fidelity and Responsibility promotes the idea that each individual within the peer mentoring relationship should have a clear understanding of their role and that these roles are all agreed upon (APA, 2012, 2017). Key to this principle is the idea of reciprocity. The peer mentoring relationship, including your role as the faculty facilitator, should be reciprocal. At no point in time should one individual use any of the other individuals for personal gain. This is especially relevant to your role as the faculty facilitator. While facilitating a peer mentoring relationship may align with expectations for service, especially as it relates to faculty evaluations, promotions, and tenure, it is essential that you facilitate the relationship in a way that also promotes goal achievement and career preparation for your mentors and mentees. Your role as a faculty facilitator should not hinge on only furthering your own career.

Integrity

The third Guiding Principle is Integrity, which aligns closely with Fidelity and Responsibility. Once roles have been established and expectations have been agreed upon, each individual within the peer mentoring relationship needs to actually follow through on what they have said they are going to do (APA, 2012, 2017). If circumstances change or problems arise, that's okay. But a conversation needs to be held where the change or problem is discussed and a clear path forward is agreed upon. As a faculty facilitator, Integrity is key to you establishing trust and rapport with your student mentors and mentees. Your students must be able to trust that you will do what you say you will do.

Justice

Next is Justice. Justice promotes the idea of fairness and a relationship free of bias to the best of one's ability (APA, 2012, 2017). This aligns with the self-reflection practices and culturally responsive practices that were discussed in Chapters 2 and 3. As a faculty facilitator, Justice requires that you remain mindful of your own personal biases, engage in self-reflection on a continuous basis, and adjust your words and actions to ensure alignment with your values. This also means ensuring that you do not "play favorites" with any of your students, making sure that all of your mentors and mentees are provided equitable attention and opportunities.

Respect for People's Rights and Dignity

Aligning with Justice, the final Guiding Principle is Respect for People's Rights and Dignity. Just as with Justice, this principle coincides with our discussions in Chapters 2 and 3 on self-reflection practices and cultural responsiveness, ensuring that differences in personal experiences are viewed through a lens of value and respect and that each individual within the peer mentoring relationship is treated as a human being with value (APA, 2012, 2017). This also aligns with our discussions earlier in this chapter on power differential. As a faculty facilitator, you will need to remain mindful of the differences between your experiences and your students' experiences, the power differential that is present, and how your words and actions seek to provide a level playing field for all individuals within the relationship.

Now that you have a strong understanding of ethics in mentoring as well as practices to support your students in navigating conflicts, in the next chapter, we will explore strategies for building community and facilitating opportunity.

Summary

In summary, in this chapter, we discussed quite a few things! Namely, we reviewed the foundations of learning theory with a specific focus on the role of observation and modeling in learning. In this chapter, we also reviewed the importance of competent leadership, which involves various actions such as critical self-reflection and strengths assessments. We explored methods for being an effective communicator, including using the five steps to empathic listening while understanding the five levels of listening (with a Level 5 being your goal as a faculty facilitator). In this chapter, we explored how to build trust as the faculty facilitator through practicing unconditional positive regard, practicing empathy, and developing emotional intelligence. We dissected the importance of recognizing the context through which the peer mentoring relationship exists, your role within the hierarchical structure of the relationship, the related power dynamics that you must be aware of and mitigate, and the importance of setting personal and professional boundaries to best support your students' development of critical competencies. Strategies for managing conflict were shared, including the use of restorative practices. And, finally, the critical component of ethical mentoring was highlighted. As we wrap up this chapter and move forward in our learning, take a few moments to consider the questions and prompts presented in the Reflect and Act section.

Reflect and Act

1 Reflect on your own experience from childhood to present day. What skills did you develop as a result of observation and modeling of those around you? How might this apply to your role as a model for your students?

2 In becoming a competent leader, regular engagement in critical self-reflection is necessary. In what ways can you make this a habit? Make a plan that will support your own self-reflection.
3 Reflect on your strengths and your areas in need of improvement. What parts of the mentoring process are you skilled at? What parts of the mentoring process might you need to seek out additional resources, training, or practice in? Consider taking a strengths assessment survey (see Chapter 7 for suggestions).
4 Assess your skill at listening. Consider the five levels of listening as you engage in conversations throughout the day. Take note of the level at which you are listening to your colleagues, students, and others. What steps can you take to move toward becoming an empathic listener (Level 5)? Practice these steps until they become more natural and habitual to you.
5 Take some time to reflect on the level of trust that your current students (whether academic students or peer mentors and peer mentees) have in you. What traits, skills, or competencies do you currently have that support the building of trust between you and your students? What could you do differently to more effectively facilitate the building of trust?
6 Consider your overall level of emotional intelligence. Pay close attention when you are interacting with others to determine how well you are able to (1) recognize and understand your own feelings, emotions, and needs; (2) understand the impact of your moods and emotions on others; (3) suspend judgement when something is shared with you that you may not agree with or may not have done had the situation been reversed; (4) display flexibility in a changing or challenging situation; (5) take others' feelings, emotions, and needs into consideration; (6) respond appropriately to others' "big feelings"; and (7) engage in culturally responsive practices. Where is your level of emotional intelligence strong? Where can your level of emotional intelligence be improved? What steps can you take to model a high level of emotional intelligence?
7 Consider the context through which you are or will be facilitating the peer mentoring relationship. Take some time to write down (journal) the various contexts that you will need to be mindful of. Revisit this narrative anytime you find that you are making assumptions or viewing the peer mentoring relationship through your own personal lens rather than the lens of others within the relationship.
8 Reflect on the power dynamics within your relationships with students as peer mentors and peer mentees. Are there areas in which you may need to be more aware of the power differential? If so, how might you address these areas to mitigate the potential detrimental effects that an imbalance in power might yield?

9 Consider your personal and professional boundaries within the context of the peer mentoring relationship. Take some time to think about where you will need to set boundaries, how, and the steps that you will take to enforce your boundaries in healthy and effective ways. See Chapter 7 for resources on boundaries and self-care. Check out a few of those resources.
10 Reflect on the concept of restorative practices for managing conflict. In what ways do you think that restorative practices can support your work in managing conflict within the peer mentoring relationship? What challenges do you perceive? Make an action plan for attending to those challenges.
11 Review the five main goals for restorative practices and the three main strategies for meeting the five main goals. Consider how you might practice using the strategies when encountering conflict in your own life—separate from facilitating peer mentoring relationships. Try the strategies out until you become comfortable with implementing them routinely.
12 Review the APA resources for ethical mentoring (see Chapter 7). What strategies will you need to implement and what boundaries will you need to set in order to engage in ethical mentoring practices?
13 How do the guidelines for ethical mentoring align with the purposes of and strategies for managing conflict through restorative practices?

References

Advancement Project. (2014). *Restorative practices: Fostering healthy relationships & promoting positive discipline in schools*. The Schott Foundation for Public Education.

Amstutz, L. S., & Mullet, J. H. (2005). *The little book of restorative discipline for schools: Teaching responsibility; creating caring climates*. Good Books.

APA. (2012). *Introduction to mentoring: A guide for mentors and mentees*. https://www.apa.org/education-career/grad/mentoring

APA. (2017). *Ethical principles of psychologists and code of conduct*. American Psychological Association. Retrieved June 11, 2025, from https://www.apa.org/ethics/code

Beaudoin, M., & Maki, K. (2021). *Mindfulness in a busy world: Lowering barriers for adults and youth to cultivate focus, emotional peace, and gratefulness*. Rowman & Littlefield.

Beech, N., & Brockbank, A. (1999). Power/knowledge and psychosocial dynamics in mentoring. *Management Learning, 30*(1), 7–24.

Clarkson, P., & Shaw, P. (1992). Human relationships at work in organisations. *Management Education and Development, 23*(1), 18–29.

Costello, B., Wachtel, J., & Wachtel, T. (2009). *The restorative practices handbook for teachers, disciplinarians and administrators*. International Institute for Restorative Practices.

Covey, S. R. (2012). *The 7 habits of highly effective people: Restoring the character ethic.* RosettaBooks LLC.

Davis, J., & Coopes, R. (2022). Our fire stories. *Journal of Awareness-Based Systems Change, 2*(2), 85–108.

Dweck, C. S. (2017). From needs to goals and representations: Foundations for a unified theory of motivation, personality, and development. *Psychological Review, 124*(6), 689–719. https://doi.org/10.1037/rev0000082

Dweck, C. S., & Yeager, D. S. (2019). Mindsets: A view from two eras. *Perspectives on Psychological Science, 14*(3), 481–496.

Ferris, S. P., & Waldron, K. (2022). Learning from senior women leaders: In their own words. In B. Cozza & C. Parnther (Eds.), *Voices from women leaders on success in higher education: Pipelines, pathways, and promotion* (pp. 81–82). Routledge.

Freire, P. (1972). *Pedagogy of the oppressed.* Penguin.

Ivey, A. E., Ivey, M. B., & Zalaquett, C. P. (2014). *Essentials of intentional interviewing: Counseling in a multicultural world.* Cengage Learning.

Jones, V., & Jones, L. (2021). *Comprehensive classroom management: Creating communities of support and solving problems* (12th ed.). Pearson.

Kovach, M. (2020). Leader influence: A research review of French & Raven's (1959) power dynamics. *The Journal of Values-Based Leadership, 13*(2), 102–112.

LePera, N. (2021). *How to do the work: Recognize your patterns, heal from your past, and create your self.* HarperCollins.

Levine, L. E., & Munsch, J. (2022). *Child development: An active learning approach* (4th ed.). Sage.

McAllister, D. J. (1995). Affect- and cognition-based trust as foundations for interpersonal cooperation in organizations. *Academy of Management Journal, 38*(1), 24–59.

Milner, H. R., Cunningham, H. B., Delale-O'Connor, L., & Kestenberg, E. G. (2019). *"These kids are out of control": Why we must reimagine "classroom management" for equity.* Corwin.

Mondisa, J., & Adams, R. S. (2022). A learning partnerships perspective of how mentors help proteges develop self-authorship. *Journal of Diversity in Higher Education, 15*(3), 337–353.

Robbins, M. (2024). *The Let Them Theory: A life-changing tool that millions of people can't stop talking about.* Hay House, LLC.

Rockinson-Szapkiw, A., Wendt, J. L., & Wade-Jaimes, K. (2020). *Navigating the peer mentoring relationship: A handbook for women and other underrepresented populations in STEM.* Kendall Hunt Publishing Company.

Rogers, C. R. (1957). The necessary and sufficient conditions of therapeutic personality change. *Journal of Counseling Psychology, 2*, 95–103.

Rogers, C. R. (1959). A theory of therapy, personality, and interpersonal relationships as developed in the client-centered framework. In S. Koch (Ed.), *Psychology: A study of a science* (Vol. 3). McGraw-Hill.

Rolfe, A. (2021). *Mentoring mindset, skills and tools.* Mentoring Works.

Villanueva, I., Gelles, L., & Di Stefano, M. (2020). Understanding ethical peer mentoring. In A. Rockinson-Szapkiw, J. L. Wendt, & K. Wade-Jaimes (Eds.), *Navigating the peer mentoring relationship: A handbook for women and other underrepresented populations in STEM* (pp. 183–191). Kendall Hunt Publishing Company.

Vygotsky, L. (1978a). Interaction between learning and development. In M. Cole, V. John-Steiner, & E. Souberman (Eds.), *Mind in society: The development of higher psychological processes* (pp. 79–91). Harvard University Press.

Vygotsky, L. (1978b). *Mind in society: The development of higher psychological processes*. Harvard University Press.
Vygotsky, L. (1986). *Thought and language*. Harvard University Press.
Wells, L. M. (2024). *There are no deficits here: Disrupting anti-Blackness in education*. Corwin.
Wilmot, W. W., & Hocker, J. L. (2001). *Interpersonal conflict* (6th ed.). McGraw-Hill.
Wolfe, S. (2022). Theoretical frameworks for the success of working mothers. In B. Cozza & C. Parnther (Eds.), *Voices from women leaders on success in higher education: Pipelines, pathways, and promotion* (pp. 30–40). Routledge.
Yazzie, R. (1994). Life comes from it: Navajo justice concepts. *New Mexico Law Review, 1*(1), 179–190.
Zachary, L. J., & Fain, L. Z. (2022). *The mentor's guide: Facilitating effective learning relationships*. Jossey-Bass.
Zehr, H. (2015). *The little book of restorative justice*. Good Books.

6
BUILDING COMMUNITY AND FACILITATING OPPORTUNITY

Introduction

As has been shared throughout this book, participation in science, technology, engineering, and mathematics (STEM) degree programs and career fields currently does not accurately represent the diverse gender, cultural, racial, and ethnic demographics of the overall population (National Science Foundation [NSF], 2023). Peer mentoring, however, has been proven to be one method of helping to encourage and support individuals who have been historically underrepresented as they navigate these fields of study (National Academies of Sciences, 2019). But peer mentoring requires a skilled and competent faculty facilitator in order for student peer mentors and peer mentees to fully reap the benefits of the peer mentoring relationship (Jones & Wendt, 2025; Wendt & Jones, 2024a, 2024b).

We've discussed throughout the book why peer mentoring is important, how it benefits students academically, socially, and psychosocially, the theoretical foundations of mentoring, the importance of recognizing and respecting personal values and identities, and the value of practices that reflect cultural responsiveness, motivation, and adoption of a growth mindset. We've discussed the various phases of mentoring as well as roles and detailed strategies for you to use to effectively facilitate a peer mentoring relationship. Now we shift our focus slightly to what you as the faculty facilitator can uniquely offer to your students that no one else can really offer in quite the same way.

As you know by now, mentoring is a developmental relationship (APA, 2012) that is meant to support learning and growth. Thus, it is

DOI: 10.4324/9781003563341-8

transformative in nature (see Chapter 5). With this in mind, as you work with your students, you will have numerous opportunities to support their growth in their respective disciplinary fields. As the faculty facilitator, you are uniquely positioned to provide your students with knowledge, networks, and opportunities that your students otherwise may not be privileged enough to experience or obtain. It is your job as the faculty facilitator to be cognizant of those unique offerings and to strategically provide them to your students in a way that is both equitable and responsive to their individual needs. One such opportunity is through the building of community. Let's explore this further.

Community

As discussed in Chapters 1 and 5, humans are social beings that learn through social interactions (Vygotsky, 1978a, 1978b, 1986; Wenger, 1998). We learn, then, through community—through observing, working with, and interacting with others. How we perceive that we belong within a particular group and how we feel that our needs are met within the group is known as *sense of community* (McMillan & Chavis, 1986; Wenger, 1998). When individuals have a strong sense of community, they feel like they belong—that they are productive, contributing members of a community that in turn engages in reciprocal actions and support (like an effective peer mentoring relationship!). Sense of community is rooted within the theory of communities of practice (Wenger, 1998). The theory of communities of practice posits that learning requires social participation as it is fundamentally a social phenomenon. A *community of practice*, then, is a community that, over time, develops around mutual wants, needs, and goals. In other words, a community of practice is a "group of people who share a concern or a passion for something they do and learn how to do it better as they interact regularly" (Wenger-Trayner & Wenger-Trayner, 2015, p. 2).

The Theory of Communities of Practice

There are four central ideas that frame the theory of Communities of Practice: humans are social beings; learning occurs through shared goals that are accomplished with competence and respect; learning occurs through active engagement; and learning produces meaning (Wenger, 1998). In a nutshell, a community of practice is a group of individuals who, large or small, informally or formally, have a shared interest (Wenger-Trayner & Wenger-Trayner, 2015). Membership within the group is defined by the individuals' commitment to the shared interest (a *domain*), which simultaneously distinguishes members of the group from non-members of the group, thus forming a *community*. In engaging with their shared interest, members join in collaboration—discussions, activities, sharing of information, and any other form of reciprocity that centers around the shared interest or goal. Together, the members interact through shared *practice*, which will look different depending on the specific shared interest and shared goals.

Within the context of the peer mentoring relationship, the faculty facilitator, peer mentor, and peer mentee(s) will have a shared interest and goal, which will be dependent on the desired outcomes agreed upon by the mentors and mentees specifically—this is considered the *domain*. The faculty facilitator, peer mentor, and peer mentee(s) will work together to share information and resources, engage collaboratively in activities and problem solving, among other collaborative activities—this is considered the *community*. Working together, the faculty facilitator, peer mentor, and peer mentee(s) are practitioners engaged in a shared *practice* (see Figure 6.1). It is important to note, however, that a community of practice can be large or small, formal or informal. We are surrounded by (and engaging with) communities of practice all the time. A peer mentoring relationship is simply *one* type of community of practice.

As a faculty facilitator, it will be essential for you to develop and foster both a community of practice—where your student peer mentors and peer mentees work collaboratively toward shared wants, needs, and goals—and a sense of community—where each member of the peer mentoring relationship feels that they belong. There are myriad ways to do this—many of which we have already discussed in Chapters 4 and 5. But here it is important for you to think a bit broader.

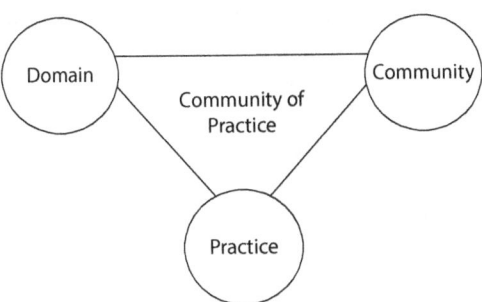

FIGURE 6.1 A community of practice is formed through a shared domain, community, and practice.

Remember how I mentioned that communities of practice can be large, small, formal, or informal and that we engage with and are members of many different communities of practice? Consider what communities of practice you are a member of. For instance, think about your professional affiliations, the professional organizations that you are a part of. Do these professional organizations offer student membership rates or even complementary or reduced cost workshops for students? What workshops, webinars, or talks can you share with students so that they can develop disciplinary knowledge and professional networks? Are there listservs that you can introduce to your students? Discipline-focused blogs? Think also about the working groups you might be a part of on campus. Think about the campus culture—research groups, writing groups, special interest groups, etc. Your students don't know what they don't know—they may not be aware that these groups and connections are even available to them. The networks that you connect students to, though, can serve as lifelines in gaining professional knowledge, making connections with like others, developing social capital, and gaining access to opportunities such as internships and even job postings. There truly is no limit. How can you help your students become part of other communities of practice and develop a sense of community when considering the communities of practice that you are part of? What shared interests or goals within your communities of practice might facilitate your students' learning and growth? How can you leverage your communities of practice to prepare your students to thrive beyond the academic institution?

Forging Paths

As a faculty facilitator, you are a leader. You are a role model. You are a guide. We discussed what your role as faculty facilitator looks like in depth in Chapter 5. But the other component that I want to focus on here is your ability to forge paths both for your students and *yourself*.

Being a faculty member can be stressful at times. If you are in a tenure-track position, there is the stress of accomplishing enough to satisfy tenure requirements within a very limited amount of time—your job depends on it. Even if you are not in a tenure-track position, most (if not all) institutions have some sort of faculty evaluation process that occurs on a yearly or bi-yearly cycle. The evaluation of your performance within your faculty position is likely dependent on a combination of teaching, scholarship, and service (see Figure 6.2). Facilitating peer mentoring relationships is one way that you can attend to all three of these areas. For instance, if you are extending your knowledge of content area expertise to the peer mentoring relationships that you are facilitating, you are engaging in *teaching*. If

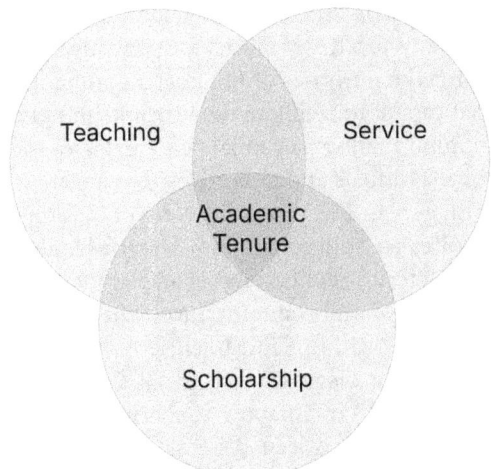

FIGURE 6.2 The intersection of teaching, scholarship, and service that typically are required to achieve academic tenure.

you are including students in research activities—whether actively doing the research, presenting the research, attending conferences, writing papers, etc.—you are engaging in *scholarship* while simultaneously facilitating your students' engagement in scholarly activities. And by sheer nature of facilitating the peer mentoring relationship—giving of your time, your energy, your expertise in order to enhance your students' experiences within and beyond the walls of the institution—you are engaging in *service*. Facilitating the peer mentoring relationship through these lenses can support your professional growth as well.

It is important to recognize, though, that you cannot do everything on your own—nor should you. It is wise to recognize your limits—both in terms of expertise and time—and lean on your connections, resources, and networks to fill in the gaps. If you are mid-career or late-career faculty, you can loop in junior faculty so that they, too, can see the benefits of facilitating peer mentoring relationships and engage in teaching, scholarship, and service. You can lean on more experienced faculty as well as staff within the institution that can provide targeted and specific support in the areas that your students are most in need. Research staff, including those who typically reside within the library, can be invaluable! Essentially, you can facilitate the building of bridges (Elgayeva, 2022), opening doors for your students, showing them how and where to find resources, while simultaneously strengthening and expanding your own professional networks. Remember that "institutions are social ecosystems" (Elgayeva, 2022, p. 43).

As a faculty facilitator, it is up to you to determine how best to navigate those ecosystems and to teach and model for your students how to navigate those ecosystems for the purposes of (1) goal attainment and (2) reciprocity. Use your social capital to facilitate the learning and growth of your students, keeping in mind that social capital "encompasses social relations that create avenues for individuals and collectives to navigate the complexity of social systems" (Elgayeva, 2022, p. 43). Networks forged through social systems not only offer an additional path to knowledge and opportunity, but they also establish feelings of belonging and support (Ferris & Waldron, 2022). This is especially important for those that have been historically marginalized, particularly within STEM fields.

Tying this to what we discussed on Social Cognitive Career Theory (SCCT) (Lent et al., 1994) in Chapter 1, sharing access to networks and facilitating the navigation of social systems is essential to effectively facilitating peer mentoring relationships. Remember that a key concept of SCCT is that persistence, motivation, and engagement facilitate opportunities for growth. And opportunities for growth require an individual to try new tasks, develop mastery, and obtain useful skills (Bandura, 1977, 1997). Tapping into your social capital through your networks is one way that you can provide your students with opportunities to try new tasks, develop mastery, and obtain useful skills by introducing your students to knowledgeable others.

Considering the structure of the peer mentoring relationship, while the peer mentor likely has more experience and more connections than the peer mentee, these two individuals are essentially near-peers. This means that there is a limit to the knowledge, expertise, and experience that they hold and, thus, can share, so they'll be looking to you to help open doors and give direction that they otherwise would not be privy to or, in some cases, may not even be aware exist. As a faculty facilitator, your role is that of a transformative leader (Ferris & Waldron, 2022; Wolfe, 2022). Transformative leadership supports the key concept of the peer mentoring relationship being reciprocal—information, resources, and opportunities are shared and exchanged amongst all individuals in the relationship (Wolfe, 2022). This exchange is collaborative in nature and dependent on the quality and strengths of the relationships that you forge with your students. In turn, the quality and strength of the relationships that you forge with your students are dependent on the sense of community that you cultivate.

It will also largely be up to you to challenge barriers that might prohibit your students from being successful and/or achieving their desired goals (Parnther, 2022)—whatever those barriers might happen to look like. It will largely be up to you to promote pathways that align with your students' stated goals. It will largely be up to you to create opportunities for your students, challenging traditional models that, historically, have seemed to almost mimic a form of indentured servitude. It will largely be up to you to acknowledge the talents, potential, and contributions of your students to advancing the body of knowledge, particularly in the fast-paced and ever-changing realm of STEM, as a form of *collective labor*. Simultaneously, it will be up to you to unapologetically set boundaries that both protect yourself and, to an appropriate extent, your students (remembering that mistakes are also part of the learning process). Importantly, it will be essential that you take these actions while also honoring your own and your students' unique and intersectional identities. Strategically making the most of your social capital will help with each of these imperatives.

Relationship Constellations

Given that each of us has experience and knowledge that will differ from the experiences and knowledge of others, it is likely (and preferable) that your students will have multiple mentors throughout the course of their studies and career (APA, 2012; Yip & Kram, 2017). This introduces a variety of perspectives and experiences and, thus, will assist your peer mentors and mentees in receiving the support and guidance that they need as they advance within their individual journeys. Within the research literature, we refer to this as *relationship constellations* (APA, 2012; Kram, 1985). Each relationship constellation provides different supports to mentors and mentees as they grow and develop. Together, these relationship constellations form a developmental network (Higgins & Kram, 2001; Yip & Kram, 2017). What is most exciting about this concept is that it means that you don't have to have all of the answers or all of the resources or even all of the connections and networks! What it means is that it is okay and actually preferred that you lean on your own relationship constellations to build bridges and make ties between your existing relationships and your mentors and mentees. You are, therefore, helping to weave your students' networks by simply putting them in contact with those that you know can help with whatever knowledge, expertise, opportunity, or resource that your students may need. You are building a community, a network, a support system. Isn't that cool?

From a theoretical perspective, the connection between developmental networks and mentoring relationships is grounded in the idea that two main components regarding relationships influence mentors' and mentees' outcomes: strength of the developmental tie and network diversity (Higgins & Kram, 2001; Yip & Kram, 2017). When mentors and mentees have strong developmental networks, career outcomes, job satisfaction, identity, and commitment are enhanced (Dobrow & Higgins, 2005; Higgins & Thomas,

2001; Murphy & Kram, 2010; Yip & Kram, 2017). The strength of a developmental network is measured by "the amount of time, the emotional intensity, the intimacy (mutual confiding), and the reciprocal services" (Granovetter, 1973, p. 1361) which characterize the relationship. When those networks also include diverse perspectives and knowledge, the quality of mentors' and mentees' development is enhanced. Simply put, when your mentors and mentees are able to depend on a high quality, diverse, yet small network of relationships to support their growth and development rather than looking to you solely to meet all of their needs, then they will experience better outcomes.

However, keep in mind that networks require a balance. I'm sure you've heard the saying that "things get messy when there are too many cooks in the kitchen." Well, there can be too many individuals within a network, too, which can introduce overwhelm and confusion. Research suggests that "there is clearly a balance to achieve between depth of relationships and breadth of relationships" (Yip & Kram, 2017, p. 9). Thus, be strategic. This is where I urge you to consider your basic role as a faculty facilitator—supporting when and as needed based on your students' goals and individual needs. Let me reiterate—you are not expected to do it all. You are but one cog in the wheel (albeit a very important cog in the wheel) that is supporting the forward momentum of your students.

Let's now move to discussing the culture, norms, and practices that you will want to help your students acclimate to and, in some cases, challenge in order to ensure your students are making progress toward their goals.

Culture, Norms, and Practices

Every discipline area, institution, and workplace has a specific culture which includes accepted norms and practices that a successful individual within that space is expected to conform to (whether this is "right" or "wrong" is another thing … we'll get to that in a minute). As Wolfe shares, "There are opportunities for career advancement for employees who understand the language and behavior that is expected within a particular work environment, and how systems reward people who adopt and adapt to its culture, norms, and practices" (p. 32). This extends beyond the workplace to students within an institution's environment, students within particular laboratory environments, and beyond. When a student does not conform to the specified culture, norms, and practices, then they are often not provided with the same opportunities for growth and potentially ostracized. At the very least, their progress toward goal attainment is made more challenging. Their sense of belonging is often imperiled and the privilege of becoming part of the community of practice is stunted. Thus, non-conformity for many is too great a cost to bear (Wolfe, 2022). Let's explore the theory that frames this understanding, Person-Environment Fit (Holland, 1966, 1973).

Person-Environment Fit

Within the research literature, the theory of Person-Environment Fit (Holland, 1966, 1973) is quite prevalently used to explain how (or whether) an individual matches or "fits" within a particular environment, be it an academic (institutional) environment, career environment, or other professional environment (Garibay, 2020; Wolfe, 2022). Person-Environment Fit is predicated on the idea that students can be categorized in one of six types depending on their individual personalities and their specific environments (realistic, investigative, artistic, social, enterprising, and conventional) (Holland, 1966, 1973). Students will seek out environments that match their preferences, personalities, interests, and abilities (see Figure 6.3). The environment within which they are socialized largely determines and

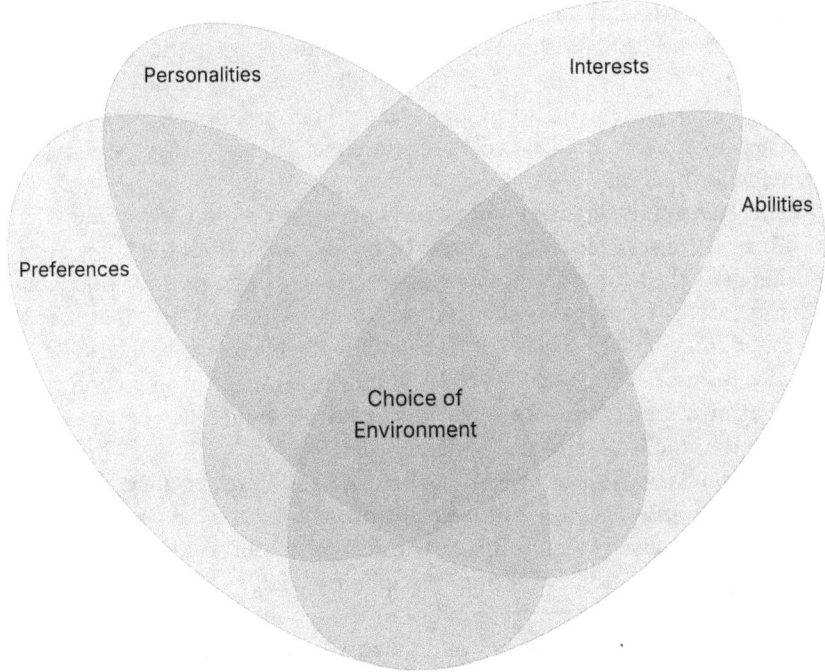

FIGURE 6.3 The environment an individual chooses is the result of the combined influence of preferences, personalities, interests, and abilities.

impacts how students acquire their preferences, personalities, interests, and abilities. Thus, students will find that they are most successful within environments that match or "fit" their personality type (the six aforementioned types). Students, then, will likely choose to engage in fields that reflect their own personality, which in turn leads to increased retention and success (Wolfe, 2022).

There are two main areas of thought on the application of Person-Environment Fit, which relate to the peer mentoring experience. The first area of thought is that students are more likely to enter and remain in fields that fit with their personality (Garibay, 2020; Smart et al., 2006). Thus, these students will be more successful than those who enter a field that does not fit their personality. This means that faculty, including faculty facilitators, should guide students into fields that match their personality type. This, however, could lead faculty facilitators into limiting students' choices unnecessarily.

The second area of thought is that when students enter into fields that match their personality, sometimes their dominant personality that most closely aligns with their field of choice remains stable and other times it becomes minimized (Garibay, 2020; Smart et al., 2006). Even if the student's dominant personality becomes minimized (or even changes), the student can still be successful within that field if they adopt the culture and norms of that specific space. This means that faculty, including faculty facilitators, should guide students to adopt the accepted culture, norms, and practices that are inherent to the field that the students wish to be successful in—essentially assimilating.

The question is, do assimilation and conformity align with students' individual identities, needs, and goals? In other words, does the approach of pushing students toward specific fields based on personality and/or encouraging students to adopt specific cultures, norms, and practices that may be in misalignment with who they are as individuals empower them (Bensimon & Bishop, 2012; Garibay, 2020)? Or does assimilation and conformity work against students in ways that detrimentally limit their opportunities and/or devalue who they are as individuals?

As a faculty facilitator, it is my belief that a delicate balance is needed. While there are expected cultures, norms, values, and practices that are inherent to each respective space within which our students operate—including within their specific disciplinary fields and future career environments—it is the responsibility of the faculty facilitator to be both cognizant of and supportive of students' individualized experiences, cultures, beliefs, values, personalities, identities, needs, and goals. As was discussed in Chapter 3, the education system (particularly within the US) has been built upon a system of inequities that perpetuates dominant expectations and beliefs (Bensimon & Bishop, 2012). The world of postsecondary education is no different. Thus, it is up to you as the faculty facilitator to ensure that you are helping your students prepare in a way that facilitates their success—that attends to the cultures, norms, and practices that they will need to navigate within their chosen field—while also remaining cognizant of, respectful of, and attentive to their own individual experiences, identities, cultures, and beliefs.

This means that sometimes you may need to push the boundaries, moving toward a more equitable environment. You may need to challenge expectations, working toward reducing the perpetuation of dominant beliefs. You may need to work in creative ways to minimize the inequities that students may experience in access and opportunities for success. In fact, I encourage you to do so—especially for your students who are marginalized. You have an obligation as a faculty facilitator to help students navigate the institutional environment in a way that aligns with your values as well as theirs. You have an obligation to serve as a role model "in practicing what [you] preach to students about how to lead fulfilling and balanced lives both professionally and personally, and show how to be cognizant of the obligations to themselves and how they define success" (Guthrie et al., 2005; Wolfe, 2022, p. 33).

Supporting Creativity

You also have an obligation as a faculty facilitator to support students' creativity. It is easy to fall into the trap of believing that you as the expert are the only person with novel ideas. But students bring a different set of experiences, a different lens, a fresh set of eyes that can often find solutions and create products in ways that are new and exciting! And even if you don't fully believe that your students' ideas or creations are inherently better than your own (move aside, ego!), letting them explore and try is essential to their own learning and growth. You were in their shoes not so long ago. After all, isn't that what facilitating a peer mentoring relationship is about? Facilitating learning and growth? It certainly isn't about your ego!

We have to let students think outside the box sometimes, and that moves beyond creating their own goals and working amongst themselves to manage challenge (see Chapters 4 and 5). As a faculty facilitator, let your students be creative. Let them explore and try new ideas. If they make a mistake, it's okay. The worst case scenario is that it will serve as a learning experience. The best case scenario is that they develop a new prototype, discover a new solution, create a masterpiece—whatever the case may be.

You are facilitating them in learning and growing toward meeting their goals, which means that you should encourage their creativity as they explore who they are within the world that they are learning to navigate. You are facilitating their moving to the "next level," which now brings me to shifts in the relationship role. Let's discuss.

Shifts in Relationship Role

As has been mentioned numerous times throughout this book, your role as a faculty facilitator is that of a transformative leader. This means that, at times, the peer mentoring relationship roles may shift and morph over time depending on experience and needs and as your students learn, grow, and become the future experts. As faculty, you may find that you were at one time teaching a particular individual in one of your classes and now have found yourself as a faculty facilitator for this same individual within a peer mentoring relationship. Or perhaps the peer mentoring relationship is ending and now you are finding yourself working as a colleague with one of your former peer mentoring students. Relationship roles shift and change often, and it is important that you and your students (1) recognize this as being part of the normal trajectory of the faculty facilitator experience and (2) respect the new roles by modifying your expectations, words, and actions to align with the new roles.

When relationship roles shift, it is essential that each individual within the relationship communicate clearly what the shift will look like. For instance, what role will each individual now be taking? What are the respective responsibilities of each new role? Have the goals of the relationship changed if the relationship is remaining intact? Or is the relationship ending and a different relationship taking its place? Once the relationship, roles, and responsibilities are identified and agreed upon, it is helpful to clearly communicate any remaining expectations. If necessary, an agreement can be drafted and signed by all relevant individuals (much like the mentoring agreement referred to in Chapter 4). This will enable all individuals involved, including yourself, to have a clear path forward. This is also a good time to consider communities of practice and relationship constellations, which enable multiple groups and individuals to work together for different needs in the ways that work best given the specific situation, context, and parties.

I am hopeful that, at this point, you have a strong understanding of how to most effectively facilitate peer mentoring relationships. In the next chapter, I'll share some of my most utilized (and favored) resources so that you can continue your learning in a way that is most meaningful to you, tailored to your specific needs.

Summary

In summary, in this chapter, we discussed the concept of community and belonging as well as communities of practice. I explained the importance of multiple sources of knowledge, including participation in relationship constellations. As a faculty facilitator, your role in forging paths for your students was shared, including sharing opportunities and networks as appropriate. Person-Environment Fit was explained, including the challenges that culture, norms, and practices often present for marginalized populations in particular. The role of the faculty facilitator in supporting students' creativity, allowing for mistakes and stumbles for the purpose of learning (when appropriate) was explored. Finally, steps for navigating shifts in relationship roles were summarized. As we wrap up this chapter and move forward in our learning, take a few moments to consider the questions and prompts presented in the Reflect and Act section.

Reflect and Act

1 Reflect on your experiences in post-secondary education, early career, and now. When did you experience a sense of community? What helped you to feel that you belonged? What hindered your sense of community? How might you use this information to facilitate your students' sense of community?
2 What communities of practice, both formal and informal, are you a part of currently? In what ways might you introduce your students to your communities of practice? How might this facilitate their goal attainment, learning, and growth?
3 What opportunities do you have to involve students in your own work to help establish communities of practice and provide opportunities? For example, what professional affiliations and networks can you introduce your students to? What conferences might your students attend? What workshops can you share with them? What research projects could you include them in? Is there a poster presentation, conference presentation, or article that they could assist you in constructing? Think outside of the box.
4 What relationship constellations are you a part of? How can you capitalize on these relationship constellations to further support students?
5 How might you encourage students to build their own relationship constellations?
6 What cultures, norms, and practices will your students need to be aware of and potentially adhere to in order to be successful in their respective discipline areas? Do these cultures, norms, and practices align with expectations of human value and equity? If not, how might you challenge

these cultures, norms, and practices as an established member of the "expert" community? How might you assist students in navigating these cultures, norms, and practices?
7 What are the benefits of Person-Environment Fit? What are the drawbacks of Person-Environment Fit that you need to be aware of and potentially mitigate?
8 Engage in self-reflection. When students propose a new idea, solution, product, etc. that is different from your own ideas (whether better or worse), how do you react? What words and actions do you use? Do these words and actions support student creativity? How might you better support students' own creativity?
9 Consider shifts in relationship role. Have you ever experienced this type of shift? If so, what was challenging about the shift? How might you assist students in navigating shifts in relationship roles effectively? How might you personally prepare for navigating shifts in relationship roles effectively?

References

APA. (2012). *Introduction to mentoring: A guide for mentors and mentees.* https://www.apa.org/education-career/grad/mentoring

Bandura, A. (1977). Self-efficacy: Toward a unifying theory of behavioral change. *Psychological Review, 84*(2), 191–215.

Bandura, A. (1997). *Self-efficacy: The exercise of control.* W. H. Freeman and Company.

Bensimon, E. M., & Bishop, R. (2012). Why "critical"? The need for new ways of knowing. *Review of Higher Education, 36*(1), 1–7.

Dobrow, S. R., & Higgins, M. C. (2005). Developmental networks and professional identity: A longitudinal study. *Career Development International, 10*(6/7), 567–583.

Elgayeva, E. (2022). A social network framework for navigating dynamic tensions in leadership development. In B. Cozza & C. Parnther (Eds.), *Voices from women leaders on success in higher education: Pipelines, pathways, and promotion* (pp. 41–54). Routledge.

Ferris, S. P., & Waldron, K. (2022). Learning from senior women leaders: In their own words. In B. Cozza & C. Parnther (Eds.), *Voices from women leaders on success in higher education: Pipelines, pathways, and promotion* (pp. 81–82). Routledge.

Garibay, J. C. (2020). Can Holland's Person-Environment Fit Theory produce troubling outcomes for racial/ethnic underrepresented students in STEM? An analysis of social agency. *Journal Committed to Social Change on Race and Ethnicity, 6*(2), 136–176.

Granovetter, M. S. (1973). The strength of weak ties. *American Journal of Sociology, 78*(6), 1360–1380.

Guthrie, V. L., Woods, E., Cusker, C., & Gregory, M. (2005). A portrait of balance: Personal and professional balance among student affairs educators. *The College Student Affairs Journal, 24*(2), 110–127.

Higgins, M. C., & Kram, K. E. (2001). Reconceptualizing mentoring at work: A developmental network perspective. *Academy of Management Review*, 26(2), 264–288.

Higgins, M. C., & Thomas, D. A. (2001). Constellations and careers: Toward understanding the effects of multiple developmental relationships. *Journal of Organizational Behavior*, 22(3), 223–247.

Holland, J. L. (1966). *The psychology of vocational choice*. Blaisdell.

Holland, J. L. (1973). *Making vocational choices: A theory of vocational personalities and work environments*. Prentice-Hall.

Jones, V. O., & Wendt, J. L. (2025). Encouraging confidence: The impact of an online peer mentoring program on women peer mentees in STEM at two HBCUs. *Trends in Higher Education*, 4(3). https://doi.org/10.3390/higheredu4010003

Kram, K. E. (1985). *Mentoring at work: Developmental relationships in organizational life*. Foresman and Company.

Lent, R., Brown, S., & Hackett, G. (1994). Toward a unifying social cognitive theory of career and academic interest, choice, and performance. *Journal of Vocational Behavior*, 45(1), 79–122.

McMillan, D. W., & Chavis, D. M. (1986). Sense of community: A definition and theory. *Journal of Community Psychology*, 14(1), 6–23.

Murphy, W., & Kram, K. E. (2010). Understanding non-work relationships in developmental networks. *Career Development International*, 15(7), 637–663.

National Academies of Sciences, Engineering, and Medicine. (2019). *The science of effective mentorship in STEMM*. The National Acadmies Press. https://doi.org/10.17226/25568

National Science Foundation. (2023). *Diversity and STEM: Women, minorities, and persons with disabilities*. National Center for Science and Engineering Statistics. https://www.nsf.gov/reports/statistics/diversity-stem-women-minorities-persons-disabilities-2023

Parnther, C. (2022, February 8). *Advancing to leadership roles in higher education: A conversation with scholars Barbara Cozza and Ceceilia Parnther* [Interview]. American Council on Education (ACE). https://www.acenet.edu/Events/Pages/Womens-Leadership-Speaker-Series.aspx

Smart, J. C., Feldman, K. A., & Ethington, C. A. (2006). *Holland's theory and patterns of college student success*. N. S. o. P. S. Success. https://nces.ed.gov/npec/pdf/smart_team_report.pdf

Vygotsky, L. (1978a). Interaction between learning and development. In M. Cole, V. John-Steiner, & E. Souberman (Eds.), *Mind in society: The development of higher psychological processes* (pp. 79–91). Harvard University Press.

Vygotsky, L. (1978b). *Mind in society: The development of higher psychological processes*. Harvard University Press.

Vygotsky, L. (1986). *Thought and language*. Harvard University Press.

Wendt, J. L., & Jones, V. O. (2024a). Peer mentors' experiences in an online STEM peer mentoring program: "Beacons of light." *International Journal of Mentoring and Coaching in Education*, 13(3). https://doi.org/10.1108/IJMCE-03-2023-0033

Wendt, J. L., & Jones, V. O. (2024b). Supporting BIPOC males in STEM: Insights from a case study on online peer mentoring. *Journal of Research in STEM Education*, 10(1–2), 89–113. https://doi.org/10.51355/j-stem.2024.145

Wenger, E. (1998). *Communities of practice: Learning, meaning, and identity*. Cambridge University Press.

Wenger-Trayner, E., & Wenger-Trayner, B. (2015). *An introduction to communities of practice: A brief overview of the concept and its uses.* https://www.wenger-trayner.com/introduction-to-communities-of-practice/

Wolfe, S. (2022). Theoretical frameworks for the success of working mothers. In B. Cozza & C. Parnther (Eds.), *Voices from women leaders on success in higher education: Pipelines, pathways, and promotion* (pp. 30–40). Routledge.

Yip, J., & Kram, K. E. (2017). Developmental networks: Enhancing the science and practice of mentoring. In D. A. Clutterbuck, F. K. Kochan, L. Lunsford, N. Dominguez, & J. Haddock-Millar (Eds.), *The SAGE handbook of mentoring* (pp. 88–104). SAGE.

PART III
Resources for Enhancing Knowledge

7
SUGGESTED RESOURCES

Introduction

Throughout this book, I have done my best to share simple strategies that you can use to become better able to support students as they engage in peer mentoring relationships as a faculty facilitator. It is my belief that faculty (as most educators) at their core want to cultivate environments and experiences for their students that are nurturing, welcoming, and supportive. Yet, faculty increasingly feel the strains of more responsibilities with limiting time constraints. Quite simply, there are not enough of us to go around, nor do we seem to ever have enough time! By capitalizing on the availability of students to engage in peer mentoring and cultivating the skills necessary to effectively and efficiently support and guide peer mentoring relationships, faculty can remain involved without becoming overburdened. It is my hope that this guide has helped you do just that.

As I shared at the beginning of the book, my goal was to present the information within this book in a simple yet meaningful way. However, you may find yourself curious about certain aspects of identity, cultural responsiveness, peer mentoring training for your students, or other areas. If so, I encourage you to take the time to further explore these areas, taking a deeper dive into the research literature and seeking additional resources that can support your growth and development. What follows is a list of the resources that I have found to be most helpful. This list is certainly not comprehensive but can serve as a strong starting point. I wish you the very best as you move forward in your journey to support your students!

DOI: 10.4324/9781003563341-10

Additional Suggested Resources

The following list of additional suggested resources is organized in alphabetical order by topic. Again, this is not an exhaustive list but should certainly help get you started in taking a deeper dive into the nuances that influence various aspects of facilitating the peer mentoring relationship.

Acquaintance Activities

Jones, V., & Jones, L. (2021). *Comprehensive classroom management: Creating communities of support and solving problems.* Pearson.

- I have adopted this text as the required text for the numerous classroom management courses that I teach. While it is geared toward K-12 educators and, of course, focuses on effective classroom management practices, pages 89–102 provide many examples of acquaintance activities that are fun, engaging, and can easily be modified to be developmentally appropriate for students enrolled in institutions of higher education. At the very least, this will provide you with a reference for getting started on brainstorming activities that might be most appropriate for your students.

Pack, J. A. (n.d.). A collection of icebreakers and connection activities. https://inclusiveschools.org/wp-content/uploads/Student_Connection_Activities_and_Icebreakers.pdf

- In this guide, written for group facilitators, Pack presents myriad acquaintance activities that can be used at various education levels and in various contexts. Specific examples are provided as well as tips for successfully constructing a welcoming and open environment.

Cross-Cultural Mentoring Inventories

The Chronicle of Evidence-Based Mentoring Cross-Cultural Mentoring Inventory. (2014). Cross-Cultural Mentoring Inventory (Revised). https://www.evidencebasedmentoring.org/wp-content/uploads/2014/11/Cross-CulturalInventoryRevised-Mentors.pdf

- This URL links to a free cross-cultural mentoring inventory that can be used to self-assess your level of comfort and flexibility in recognizing and attending to cultural differences.

Undergraduate Research Committee, New York City College of Technology. (n.d.). *A handbook on mentoring students in undergraduate research:*

Proven strategies for success. Press Books. https://pressbooks.cuny.edu/undergraduateresearchmentoringhandbook/

- This handbook provides guidance on mentoring students specifically within research at the undergraduate level. Importantly, a brief inventory for cross-cultural mentoring is provided at the end of Section 3.4.

Zachary, L. J., & Fain, L. Z. (2022). *The mentor's guide: Facilitating effective learning relationships.* Jossey-Bass.

- This guidebook focuses on the work inherent to mentoring from the mentor's perspective, with the grounding understanding that effective mentoring requires (and leads to) growth for both the mentor and mentee. Through practical steps, thoughtful questions, and reproducible worksheets, this resource is a valuable addition to enhancing your understanding of the mentoring relationship. In particular, I have found the resources on cross-cultural mentoring incredibly helpful (see page 43 of Zachary and Fain's work for a Cross-Cultural Mentoring Skills Inventory). (See also Mentoring Training, Tools, and Guides.)

Cultural Responsiveness, Anti-Racism, and Resources on Cultivating Inclusivity

Applebaum, B. (2022). *White educators negotiating complicity: Roadblocks paved with good intentions.* Lexington Books.

- This book is quickly becoming one of my go-to texts on understanding the role of White folks' complicity in perpetuating systemic issues with educational equity—even if unintentional. As a White educator, I believe it is incredibly important for us to mindfully confront the historical contexts that have shaped the education landscape in the US, which includes (but of course is not limited to) racial inequities. This book is eye-opening.

Hammond, Z. (2015). *Culturally responsive teaching & the brain: Promoting authentic engagement and rigor among culturally and linguistically diverse students.* Corwin.

- This resource is a text that is often used in teacher education programs to prepare teachers for creating learning environments and learning experiences that are culturally responsive. As the title implies, Hammond's focus on the brain-based components of culturally responsive learning will leave you with an enhanced understanding of not only *how* to engage in culturally responsive teaching, but *why* you should do so.

Hubbard, F. (2024). *The equity expression: Six entry points for nonnegotiable academic success.* Corwin Press, Inc.

- While written primarily for the K-12 educator, this book is just as applicable to faculty in higher education. Using a workbook-like approach with numerous opportunities for engaging with content and participating in critical reflection, this is a wonderful resource for helping you interrogate your own unconscious biases, how your biases might be evident within your actions and behaviors, and steps to take to engage in more equitable support for your students.

Ladson-Billings, G. (2021). *Culturally relevant pedagogy: Asking a different question.* Teachers College Press.

- In this book, Gloria Ladson-Billings—a renowned expert in the field of education—outlines what it means to be a culturally relevant and responsive educator. Ladson-Billings provides context, theory, and tangible steps for establishing a culturally responsive learning experience in a manner that is accessible to those with an education background as well as those that are not. In my opinion, this should be required reading for all educators.

Ladson-Billings, G. (2005). Is the team alright? Diversity and teacher education. *Journal of Teacher Education, 56*(3), 229–234.

- I recommend that all faculty—particularly those of us that are White—read this article. While focused on teacher education, Ladson-Billings sheds light on the mismatch between students' identities and faculty members' identities, particularly racial and ethnic identities. She does a brilliant job of highlighting several of the pitfalls that White faculty tend to fall into—each of which we need to be aware of in order to do a better job welcoming those with different identities and experiences than our own.

Miller, E. T., & Walker, A. V. (2023). *Antiracist pedagogy in action: Curriculum development from the field.* Rowman & Littlefield Publishing.

- From the viewpoint of a group of college instructors, K-12 teachers, and a program director, this edited collection provides much-needed insight on the role of antiracist pedagogy in a democratic society—specifically, a democratic education. With first-hand examples and supported robustly with citations from the research literature, the insights derived from this book can be applied both in and outside of the classroom to ensure that all students are provided opportunities to feel welcome, included, and able to learn.

Moore, E., Michael, A., & Penick-Parks, M. W. (2018). *The guide for White women who teach Black boys*. Corwin.

- As the title suggests, this book is written primarily to support White women—who make up the largest proportion of the teaching workforce in the US—in engaging in culturally responsive, equitable, and justice-driven practices to support Black boys as well as other historically marginalized students. The lessons in this book are easily applicable to facilitating the peer mentoring relationship as they provide context to explain the current state of the education system in the US and the challenges that minoritized populations experience within the US education system, which, of course, do not end at high school graduation. Chapters span topics such as identity, culturally responsive practices, managing conflict, and relationship building. This is a fantastic resource for anyone that seeks to improve their teaching practices.

Wells, L. M. (2024). *There are no deficits here: Disrupting anti-Blackness in education*. Corwin.

- This is a fantastic resource for understanding the impact of the historically perpetuated deficit-minded approach that the US education system by and large has taken. In this book, Wells focuses on anti-Blackness, but I would argue that their work is just as applicable to other marginalized populations. Of specific interest, I would encourage you to explore Chapter 9 of Wells's book that focuses on culturally-responsive sustaining education.

Emotional Intelligence

Yale School of Medicine. (2025). Yale Center for Emotional Intelligence. https://medicine.yale.edu/childstudy/services/community-and-schools-programs/center-for-emotional-intelligence/

- The Yale Center for Emotional Intelligence, housed within the Yale School of Medicine, focuses its work on the core belief that emotions matter. Their goal is to support educators—including faculty—in understanding the importance and value of emotions, learning and teaching emotional intelligence skills, and supporting positive school climates. The Center's website presents research and opportunities for professional development (including training and courses) for developing emotional intelligence-related competencies.

Empathic Listening

Covey, S. R. (2012). *The 7 habits of highly effective people: Restoring the character ethic.* RosettaBooks LLC.

- This book is a valuable resource for learning more about being effective and efficient overall. However, I've found it especially useful in learning more about enhancing my own motivation, communication, and (perhaps most importantly) skills with empathic listening. I strongly recommend this book to anyone interested in becoming better at managing all of the things that life presents, but especially in becoming a more skilled listener.

Ethical Mentoring

American Psychological Association (APA). (2012). *Introduction to mentoring: A guide for mentors and mentees.* APA. https://www.apa.org/education-career/grad/mentoring

- This guide, published by the American Psychological Association (APA), outlines many aspects of the mentoring relationship from definitions to structures to contexts. Importantly, though, an entire section is devoted to ethics in the mentoring relationship, which presents the APA *Ethical Principles of Psychologists and Code of Conduct* within the particular context of mentoring. This is my go-to guide for making sure that my words and actions as a faculty facilitator are in alignment with expectations for ethical mentoring.

American Psychological Association (APA). (2017). *Ethical principles of psychologists and code of conduct.* APA. https://www.apa.org/ethics/code

- This page is where the American Psychological Association (APA) *Ethical Principles of Psychologists and Code of Conduct* is published and includes the most recent updates. In alignment with the APA *Introduction to Mentoring: A Guide for Mentors and Mentees*, this outlines the Guiding Principles, which should scaffold your interactions with your student peer mentors and mentees to ensure ethical mentoring practices.

Villanueva, I., Gelles, L., & Di Stefano, M. (2020). Understanding ethical peer mentoring. In A. Rockinson-Szapkiw, J. L. Wendt, & K. Wade-Jaimes (Eds.), *Navigating the peer mentoring relationship: A handbook for women and other underrepresented populations in STEM* (pp. 183–191). Kendall Hunt Publishing Company.

- This book chapter focuses entirely on ethical mentoring within the context of peer mentoring relationships. In it, Villanueva, Gelles, and Di Stefano present an alternate version of principles to guide ethical peer mentoring, including Beneficence, Nonmaleficence, Autonomy, Fidelity, Fairness, and Privacy that align closely with the APA Guiding Principles. Detailed scenarios and case analyses are provided, as well as a checklist for interrogating power differentials. Importantly, the information in this chapter is also intentionally focused on the STEM peer mentoring relationship.

Growth Mindset

Dweck, C. S. (2006). *Mindset: The new psychology of success.* Random House Publishing Group.

- This is my "go-to" book for understanding the concept of (and benefits of) the growth mindset. Dweck provides countless examples of why and how adopting a growth mindset facilitates learning, growth, and success, and does so in a way that is accessible to even those that are not in academia. Chapters explore methods for encouraging growth mindset in various contexts, including sports, education, parenting, and business. If I were to recommend one resource for mindset, this would be it!

Dweck, C. S., & Yeager, D. S. (2019). Mindsets: A view from two eras. *Perspectives on Psychological Science, 14*(3), 481–496.

- In this article, the basic tenets of growth mindset are described, including the two types of mindsets and the resulting behaviors and actions likely to result from each mindset. An overview of the research literature on growth mindset is presented, including specific actions to support growth mindset and recommendations for future study. If you are looking for a quick overview of the growth mindset literature, this is a wonderful place to start.

Elliott-Moskwa, E. S. (2022). *The growth mindset workbook: CBT skills to help you build resilience, increase confidence & thrive through life's challenges.* New Harbinger Publications, Inc.

- Just as the name implies, this resource is a workbook to help you hone your ability to adopt and sustain a growth mindset. The workbook is full of activities, resources, and actionable steps to support you in the lifelong journey of embracing a growth mindset—allowing you to thrive! If you are interested in personal work related to growth mindset or perhaps are wanting a resource to support your ability to foster a growth mindset among others (such as students), this is the resource for you!

Historical Context of the Educational Landscape

Givens, J. R. (2021). *Fugitive pedagogy: Carter G. Woodson and the art of black teaching.* Harvard University Press.

- I had the pleasure of listening to Givens speak at an AAC&U Keystroke Writing Institute retreat while introducing this book. As an interdisciplinary educator and historian, Givens presents a detailed view of the Black experience in the United States' education system and the challenges presented by race and racism based on historical records and first-person accounts. It is my belief that supporting students' myriad identities requires an in-depth understanding of the historical context of *all* groups. As an educator at an historically Black institution, I have found this book invaluable.

Love, B. L. (2019). *We want to do more than survive: Abolitionist teaching and the pursuit of educational freedom.* Beacon Press.

- In this book, Love presents a clear and pointed view of the current educational landscape. Importantly, Love presents a historical view of the United States' education system, the impact the system has on individuals with diverse identities, and the relationship between identity, socioeconomics, opportunity, and societal structure. To say my eyes were opened by reading this book is truly an understatement. I could not recommend this book more for anyone seeking to have a better understanding of what it means to desire educational opportunity and, importantly, freedom.

Morris, M. W. (2016). *Pushout: The criminalization of Black girls in schools.* The New Press.

- Within the historical presentation of the landscape of education in the US through current times, I have found a dearth in resources that explore the experience of Black girls. This book helps fill that gap, providing an in-depth analysis of the inequitable treatment imposed on Black girls in the current education system. This resource is very important if we as educators are to fully understand the context within which academe operates and how best to support our students.

Powell, S. D. (2012). *Your introduction to education: Explorations in teaching.* Pearson.

- This textbook was my first introduction to a "more accurate" history of education within the U.S. I found the presentation to be straightforward, engaging, and enlightening. If you are interested in learning about the historical context of education from colonial times forward, this is a good place to start.

Identity and Identity Development

McLean, K. C., & Syed, M. (2014). *The Oxford handbook of identity development*. Oxford University Press, Inc.

- This handbook is a wonderful resource for anyone interested in learning more about current understanding of how individuals develop identity. Theoretical foundations for identity theory are presented, as well as a comprehensive overview of how identity is developed. The impact of culture on identity is explored as well as challenges that hinder identity development. I strongly recommend this resource for those interested in developing a robust understanding of identity and its various aspects.

Miller, D. L. (2021). *Honoring identities*. Rowman and Littlefield.

- I struggled with whether to place this resource under cultural responsiveness or identity. Alas, here we are. Honestly, this book attends to both. Within the context of respect for and honoring of diverse identities, Miller's work provides context, suggested curricular resources, and actionable steps to cultivate a welcoming, safe environment for all students. Although primarily written to attend to identity and cultural responsiveness in the classroom, the practices outlined in this book are easily transferable to the peer mentoring context.

Leadership

Cozza, B., & Parnther, C. (2022). *Voices from women leaders on success in higher education: Pipelines, pathways, and promotion*. Routledge.

- This book is an edited collection that focuses on the experiences of women in leadership, specifically within the context of higher education. I have found the lessons in this book to be incredibly relevant to the mentoring context, though, as mentoring relationships benefit greatly from a transformative leadership style given the reciprocal nature of relationship. As it is, women tend to default to a more transformative style than men. Anyone that is interested in learning how to best support those that are historically underrepresented, under-supported, under-promoted, and generally overlooked will find the discussions of theory, research, and practical applications found within this work to be priceless! I could not recommend this book more!

Rennison, C., & Bonomi, A. (2020). *Women leading change in academia: Breaking the glass ceiling, cliff, and slipper*. Cognella.

- This edited collection is focused on how women experience barriers and the specific steps that they have taken to enact change that supports more equitable experiences in the future. Chapters range from power differential to intersectionality to gender, race, and ethnicity bias to networking. This book seems to touch on a little bit of pretty much every topic explored in this current book! I strongly recommend Rennison and Bonomi's work for further reading.

Wieseman, K. C., & Weinburgh, M. H. (2009). *Women's experiences in leadership in K-16 science education communities: Becoming and being.* Springer.

- As this edited collection is focused on women leaders specifically within science education, I have found it to be incredibly helpful in understanding the challenges that women experience as they navigate the field of education within a discipline (science) historically dominated by men. First-hand accounts are provided, which include vivid descriptions of what women have done to mitigate the barriers that they face, achieve their goals, and attain success. This is a great read for women particularly within STEM.

Mentoring Training, Tools, and Guides

eSTEM Peer Mentoring. (2025). eSTEM peer mentoring: Ensuring equity in STEM. https://estemmentoring.com/

- This is the official program website for the eSTEM Peer Mentoring Program, which I collaboratively created. On this website, access is provided to a series of online peer mentor and peer mentee training modules that were constructed and tested to provide intentional training for students interested in becoming peer mentors or peer mentees. The website also includes resources that coincide with the training, such as handouts, worksheets, and related scholarly publications.

National Academies of Sciences, Engineering, and Medicine (NASEM). (2019). The Science of Effective Mentorship in STEMM Online Guide V1.0. https://nap.nationalacademies.org/resource/25568/interactive/tools-and-resources.html

- Based on the National Academies of Sciences, Engineering, and Medicine (NASEM) Science of Effective Mentorship in STEMM report, this website includes resources to support administrators, facilitators, mentors, and mentees in implementing effective mentoring relationships. Possible actions for all stakeholders for addressing common barriers are provided, as well as resources for constructing Individual Development Plans (IDPs), mentoring agreements (compacts), and mentoring plans.

Packard, B. W. (2016). *Successful STEM mentoring initiatives for underrepresented students: A research-based guide for faculty and administrators.* Stylus Publishing, LLC.

- As a fellow expert in the field of STEM mentoring, I cannot recommend Packard's book more! In this book, she provides a clear, organized, and step-by-step guide for faculty and administrators to support STEM students within the mentoring context. I have found the reader questions to be especially beneficial for engaging in learning and self-reflection.

Rockinson-Szapkiw, A. J., Wendt, J. L., & Wade-Jaimes, K. S. (2020). *Navigating the peer mentoring relationship: A handbook for women and other underrepresented populations in STEM.* Kendall Hunt Publishing Company.

- This resource is a workbook that I co-authored and co-edited to complement a series of online peer mentor and peer mentee training modules that were constructed to provide intentional training for students interested in becoming peer mentors or peer mentees (see the eSTEM Peer Mentoring resource listed prior). The workbook highlights the voices of many diverse scholars and serves as a step-by-step guide for supporting students in developing peer mentoring skills and competencies.

Rolfe, A. (2021). *Mentoring mindset, skills and tools.* Mentoring Works.

- This book is a comprehensive guide to mentoring. It outlines basic definitions of mentoring, types of mentoring practices, expectations, roles and responsibilities, the mentoring process, and mentoring skills. Importantly, Part V of the book provides step-by-step tools and guides for the various activities inherent to mentoring. This is a great guide if you want a concise and simple overview of actionable steps that can be applied to facilitating the peer mentoring relationship.

University of New Mexico. (2025). UNM Mentoring Institute. https://mentor.unm.edu/

- The University of New Mexico (UNM) Mentoring Institute offers an annual conference that is focused solely on mentoring. Their website is packed with resources to support faculty and students in all aspects of mentoring. They also offer a newsletter and frequent webinars to support the development of mentoring knowledge, skills, and competencies. The UNM Mentoring Institute even offers Mentoring Micro-Credentials!

University of Wisconsin System. (2025). Mentoring and mentorship. https://intranet.med.wisc.edu/faculty-affairs-and-development/faculty-central-resources/preparing-for-promotion/mentoring-and-mentorship/

- The University of Wisconsin-Madison has produced a website full of resources on the mentoring relationship, focusing on the perspective of both mentors and mentees. This website includes video resources to reiterate the importance of finding match or "fit" between mentor and mentee, as well as mentoring guides and worksheets to help navigate a successful mentoring relationship. This is a great resource for additional tools that you can use to support your students as they engage in the peer mentoring relationship. This is also a great resource for you to engage in mentoring relationships other than peer mentoring, including those that are situated within the research laboratory.

Zachary, L. J., & Fain, L. Z. (2022). *The mentor's guide: Facilitating effective learning relationships.* Jossey-Bass.

- This guidebook focuses on the work inherent to mentoring from the mentor's perspective, with the grounding understanding that effective mentoring requires (and leads to) growth for both the mentor and mentee. Through practical steps, thoughtful questions, and reproducible worksheets, this resource is a valuable addition to enhancing your understanding of the mentoring relationship. In particular, I have found the resources on context incredibly helpful (see Chapter 2 of Zachary and Fain's work).

Motivation

Eccles, J. S., & Wigfield, A. (2020). From Expectancy-Value Theory to Situated Expectancy-Value Theory: A developmental, social cognitive, and sociocultural perspective on motivation. *Contemporary Educational Psychology, 51.*

- In this article, the authors describe various iterations of Expectancy-Value Theory and how they have come to propose Situated Expectancy-Value Theory. The article includes a very nice diagram that can help you visualize the interrelationship between intrinsic value, motivation, engagement, and persistence. I particularly find this theory helpful in understanding the peer mentoring relationship as it directly relates students' enjoyment of particular tasks to their levels of motivation and persistence.

The National Academies of Science, Engineering, and Medicine (NASEM). (2019). *Science and engineering for grades 6–12: Investigation and design at the center.* NASEM.

- While this resource is focused on grades 6–12, it is a very good resource for understanding the influence of motivation on interest in science and engineering, learning about contemporary motivation theory, and discovering methods for increasing students' motivation in STEM fields at all levels. See Chapter 3 in particular, which focuses on various theories of motivation.

Robbins, M. (2017). *The 5 Second Rule: Transform your life, work, and confidence with everyday courage.* Savio Republic.

- In this book, Robbins presents the 5 Second Rule—a simple, effective method for increasing your personal motivation that can also be used to support and facilitate the motivation of your students within the peer mentoring relationship. In an easy-to-read and engaging format, you'll find the book full of resources to support the effectiveness of the rule as a motivational strategy, which can be applied in myriad areas of your life, both personal and professional. I highly recommend this resource.

Tinto, V. (2017). Through the eyes of students. *Journal of College Student Retention: Research, Theory, & Practice, 19*(3), 254–269. DOI: 10.1177/1521025115621917

- If I had to pick a favorite theorist based on relevance of the theory alone, Tinto would be my choice. I have personally found his work to be the most relevant to the peer mentoring process. In this article, Tinto provides a summary of theories surrounding student retention and persistence. He delineates the difference between the perspective of the student compared to the perspective of the institution, highlighting the need to pay closer attention to the student perspective if institutions are truly interested in supporting the student.

Weiner, B., Van Lange, P., Kruglanski, A., & Higgins, E. (2012). An attribution theory of motivation. In P. A. M. van Lange, A. W. Kruglanski, & E. T. Higgins (Eds.), *Handbook of theories of social psychology*, (Vol. 1, pp. 135–155). Sage. https://doi.org/10.4135/9781446249215.n8

- In this book chapter, a very nuanced description of how the Attribution Theory of Motivation was constructed is presented. If you want to better understand the various underlying theories as well as the thought process and testing of various constructs through which Weiner navigated in the realization of the theory, this is the resource for you. It is quite detailed, but worth the read!

Power Dynamics

Nic, B., & Brockbank, A. (1999). Power/knowledge and psychosocial dynamics in mentoring. *Management Learning, 30*(1), 7–24.

- In this article, Nic and Brockbank present a historical summary of the research on power, power dynamics, and psychosocial dynamics with a specific focus on how it all applies to the mentoring relationship. If you are wanting a clearly presented overview of power—especially as it relates to mentoring—this resource is a great place to start.

Restorative Practices

Costello, B., Wachtel, J., & Wachtel, T. (2019). *The restorative practices handbook: For teachers, disciplinarians and administrators.* International Institute for Restorative Practices.

- This book is a wonderful resource for learning about the theory that guides restorative practices as well as how and why restorative practices support the building of strong communities. Situated within the context of K-12 education, Costello, Wachtel, and Wachtel provide guidance as well on how to use restorative practices with adults—thus, incredibly useful for assisting with managing conflict within the peer mentoring relationship.

Milner, R. H., Cunningham, H. B., Delale-O'Connor, L., & Kestenberg, E. G. (2019). *"These kids are out of control": Why we must reimagine "classroom management" for equity.* Corwin.

- This is one of the books that I routinely require in the classroom management courses that I teach in teacher preparation programs. Written through the lens of equity and justice within the PreK-12 arena, this book not only provides historical context to explain the challenges of the current education system in the US, but also provides targeted instruction for how to cultivate an environment where all are valued and all can learn. The sections on restorative practices are particularly relevant to facilitating peer mentoring relationships, even at the post-secondary level, and managing conflict in ways that promote dialogue, empathy, and collaborative problem-solving.

Zehr, H. (2015). *The little book of restorative justice.* Good Books.

- Zehr, commonly known as the father of restorative justice practices, emphasizes the value of utilizing restorative practices within the context of justice and healing when a crime has been committed. This book will provide a good foundation for what restorative justice looks like and how it informs today's implementation of restorative practices, especially within the education system. A quick and easy read, this book is recommended to help you gain a foundational knowledge that will inform your ability to effectively manage challenges within peer mentoring relationships.

Self-Care and Boundaries

Beaudoin, M., & Maki, K. (2021). *Mindfulness in a busy world: Lowering barriers for adults and youth to cultivate focus, emotional peace, and gratefulness.* Rowman & Littlefield.

- Mindfulness has been a practice that has supported focus, peace, and gratitude for generations. While it has recently gained more attention—especially within the field of education with the inclusion of socioemotional learning—it goes without saying that engaging in mindfulness is a highly effective method for practicing self-care. In this book, Beaudoin and Maki provide research-based, actionable steps for practicing self-care as well as facilitating self-care among others—which can be easily applied to the peer mentoring relationship.

LePera, N. (2021). *How to do the work: Recognize your patterns, heal from your past, and create your self.* HarperCollins.

- This is hands-down one of my favorite resources on self-care. The insights that LePera provides are universal, clear, relatable, and engaging. She shares with transparency her own struggles and provides actionable steps on how to support your own healing and self-care. While this book has helped me immensely in my personal life, its lessons are easily applied to the professional realm as well—especially as they relate to setting and maintaining boundaries. I highly recommend this book for anyone interested in personal or professional reflection and growth.

Nordell, B. B. (2021). *Avoiding burnout: How exemplary teachers find fuel and cultivate success.* Rowman & Littlefield.

- Just as the title of this book implies, this book is all about the critical skill of self-care in order to prevent burnout—a very real and potentially detrimental condition that is not nearly as recognized within the US workforce as I personally think that it should be. This is a fantastic resource for faculty (and any educator, really) who are wanting to cultivate their own mental wellness as well as those who seek to support their students in doing the same. Chock full of first-hand accounts, questions for reflections, and steps for real-time practice, I could not recommend this resource more!

Robbins, M. (2024). *The Let Them Theory: A life-changing tool that millions of people can't stop talking about.* Hay House, LLC.

- This book has literally been a game changer in both my personal and professional life. It lives up to the hype surrounding it in popular media these days, focusing on the simple idea that adult human beings are allowed to have their own thoughts and feelings—and you are not responsible for them. You, however, are responsible for your reactions and your "next steps." This idea may seem obvious, but it is one that we often neglect to recognize—one that will save us from expending time, energy, and focus on things that we cannot control. Robbins's presentation within the book is clear, compelling, and relatable. The lessons that she shares will pay dividends in ensuring that you establish a strong balance between what you can control and what you can't (both personally and professionally) as well as reinforcing the idea that, within peer mentoring relationships, you are meant to guide and facilitate. You are not meant to have all of the answers, manage or control others' emotions or reactions, or tell others what to do. Read it. You can thank me later.

Strengths Assessment

Gallup. (2025). CliftonStrengths. https://www.gallup.com/cliftonstrengths/en/252137/home.aspx

- The CliftonStrengths assessment (formerly known as Gallup's StrengthsFinder) is perhaps one of the most widely known strengths assessments within the workplace. It is an hour-long assessment that seeks to assist you in better understanding your strengths and how you can use those strengths to support your engagement with others professionally. It is a paid assessment, but one that I have found highly worthwhile.

High5Test. (2025). High5 Test. https://high5test.com/

- High5 is a strengths assessment that is comparable to the CliftonStrengths assessment. It is a 20-minute assessment that helps you to better understand your personal motivations, what energizes you, and what gives meaning to your life. As of the printing of this book, it exists as a free assessment. I have found this assessment beneficial in understanding what drives me from both a personal and professional standpoint.

Summary

In this chapter, I have presented multiple sources to assist you in developing a more in-depth understanding of the various topics discussed throughout this book. While there are many resources that could have been presented, those that are listed are the ones that I have found to be most helpful as I have learned (through good old-fashioned trial and error as well as formal research) to be an effective faculty facilitator. As we wrap up this chapter (and the book) and move forward in our learning, take a few moments to consider the questions and prompts presented in the Reflect and Act section.

Reflect and Act

1 Which topics do you desire to learn more about? Take some time to explore the resources listed.
2 Are there topics that are not listed within this chapter that you feel you could benefit from learning more about? If so, make an action plan for seeking out additional information.
3 How might you apply the information that you have learned in this chapter and throughout the book to your work with students, particularly within the context of facilitating peer mentoring relationships?

References

American Psychological Association (APA). (2012). *Introduction to mentoring: A guide for mentors and mentees*. APA. https://www.apa.org/education-career/grad/mentoring

American Psychological Association (APA). (2017). *Ethical principles of psychologists and code of conduct*. APA. https://www.apa.org/ethics/code

Applebaum, B. (2022). *White educators negotiating complicity: Roadblocks paved with good intentions*. Lexington Books.

Beaudoin, M., & Maki, K. (2021). *Mindfulness in a busy world: Lowering barriers for adults and youth to cultivate focus, emotional peace, and gratefulness*. Rowman & Littlefield Publishing.

Costello, B., Wachtel, J., & Wachtel, T. (2019). *The restorative practices handbook: For teachers, disciplinarians and administrators.* International Institute for Restorative Practices.

Covey, S. R. (2012). *The 7 habits of highly effective people: Restoring the character ethic.* RosettaBooks LLC.

Cozza, B., & Parnther, C. (2022). *Voices from women leaders on success in higher education: Pipelines, pathways, and promotion.* Routledge.

Dweck, C. S. (2006). *Mindset: The new psychology of success.* Random House Publishing Group.

Dweck, C. S., & Yeager, D. S. (2019). Mindsets: A view from two eras. *Perspectives on Psychological Science, 14*(3), 481–496.

Eccles, J. S., & Wigfield, A. (2020). From Expectancy-Value Theory to Situated Expectancy-Value Theory: A developmental, social cognitive, and sociocultural perspective on motivation. *Contemporary Educational Psychology, 51,* 101859. https://doi.org/10.1016/j.cedpsych.2020.101859

Elliott-Moskwa, E. S. (2022). *The growth mindset workbook: CBT skills to help you build resilience, increase confidence & thrive through life's challenges.* New Harbinger Publications, Inc.

eSTEM Peer Mentoring. (2025). eSTEM peer mentoring: Ensuring equity in STEM. https://estemmentoring.com/

Gallup. (2025). CliftonStrengths. https://www.gallup.com/cliftonstrengths/en/252137/home.aspx

Givens, J. R. (2021). *Fugitive pedagogy: Carter G. Woodson and the art of Black teaching.* Harvard University Press.

Hammond, Z. (2015). *Culturally responsive teaching & the brain: Promoting authentic engagement and rigor among culturally and linguistically diverse students.* Corwin.

High5Test. (2025). High5 Test. https://high5test.com/

Hubbard, F. (2024). *The equity expression: Six entry points for nonnegotiable academic success.* Corwin Press, Inc.

Jones, V., & Jones, L. (2021). *Comprehensive classroom management: Creating communities of support and solving problems.* Pearson.

Ladson-Billings, G. (2005). Is the team alright? Diversity and teacher education. *Journal of Teacher Education, 56*(3), 229–234.

Ladson-Billings, G. (2021). *Culturally relevant pedagogy: Asking a different question.* Teachers College Press.

LePera, N. (2021). *How to do the work: Recognize your patterns, heal from your past, and create your self.* HarperCollins.

Love, B. L. (2019). *We want to do more than survive: Abolitionist teaching and the pursuit of educational freedom.* Beacon Press.

McLean, K. C., & Syed, M. (2014). *The Oxford handbook of identity development.* Oxford University Press, Inc.

Miller, D. L. (2021). *Honoring identities.* Rowman & Littlefield Publishing.

Miller, E. T., & Walker, A. V. (2023). *Antiracist pedagogy in action: Curriculum development from the field.* Rowman & Littlefield Publishing.

Milner, R. H., Cunningham, H. B., Delale-O'Connor, L., & Kestenberg, E. G. (2019). *"These kids are out of control": Why we must reimagine "classroom management" for equity.* Corwin.

Moore, E., Michael, A., & Penick-Parks, M. W. (2018). *The guide for White women who teach Black boys.* Corwin.

Morris, M. W. (2016). *Pushout: The criminalization of Black girls in schools.* The New Press.

National Academies of Sciences, Engineering, and Medicine (NASEM). (2019). The Science of Effective Mentorship in STEMM Online Guide V1.0. https://nap.nationalacademies.org/resource/25568/interactive/tools-and-resources.html

Nic, B., & Brockbank, A. (1999). Power/knowledge and psychosocial dynamics in mentoring. *Management Learning*, 30(1), 7–24.

Nordell, B. B. (2021). *Avoiding burnout: How exemplary teachers find fuel and cultivate success*. Rowman & Littlefield Publishing.

Pack, J. A. (n.d.). A collection of icebreakers and connection activities. https://inclusiveschools.org/wp-content/uploads/Student_Connection_Activities_and_Icebreakers.pdf

Packard, B. W. (2016). *Successful STEM mentoring initiatives for underrepresented students: A research-based guide for faculty and administrators*. Stylus Publishing, LLC.

Powell, S. D. (2012). *Your introduction to education: Explorations in teaching*. Pearson.

Rennison, C., & Bonomi, A. (2020). *Women leading change in academia: Breaking the glass ceiling, cliff, and slipper*. Cognella.

Robbins, M. (2017). *The 5 Second Rule: Transform your life, work, and confidence with everyday courage*. Savio Republic.

Robbins, M. (2024). *The Let Them Theory: A life-changing tool that millions of people can't stop talking about*. Hay House, LLC.

Rockinson-Szapkiw, A. J., Wendt, J. L., & Wade-Jaimes, K. S. (2020). *Navigating the peer mentoring relationship: A handbook for women and other underrepresented populations in STEM*. Kendall Hunt Publishing Company.

Rolfe, A. (2021). *Mentoring mindset, skills and tools*. Mentoring Works.

The Chronicle of Evidence-Based Mentoring Cross-Cultural Mentoring Inventory. (2014). Cross-Cultural Mentoring Inventory (Revised). https://www.evidence-basedmentoring.org/wp-content/uploads/2014/11/Cross-CulturalInventoryRevised-Mentors.pdf

The National Academies of Science, Engineering, and Medicine (2019). *Science and engineering for grades 6–12: Investigation and design at the center*. NASEM.

Tinto, V. (2017). Through the eyes of students. *Journal of College Student Retention: Research, Theory, & Practice*, 19(3), 254–269. DOI: 10.1177/1521025115621917

Undergraduate Research Committee, New York City College of Technology. (n.d.). *A handbook on mentoring students in undergraduate research: Proven strategies for success*. Press Books. https://pressbooks.cuny.edu/undergraduateresearchmentoringhandbook/

University of New Mexico. (2025). UNM Mentoring Institute. https://mentor.unm.edu/

University of Wisconsin System. (2025). Mentoring and mentorship. https://intranet.med.wisc.edu/faculty-affairs-and-development/faculty-central-resources/preparing-for-promotion/mentoring-and-mentorship/

Villanueva, I., Gelles, L., & Di Stefano, M. (2020). Understanding ethical peer mentoring. In A. Rockinson-Szapkiw, J. L. Wendt, & K. Wade-Jaimes (Eds.), *Navigating the peer mentoring relationship: A handbook for women and other underrepresented populations in STEM* (pp. 183–191). Kendall Hunt Publishing Company.

Weiner, B., Van Lange, P., Kruglanski, A., & Higgins, E. (2012). An attribution theory of motivation. In P. A. M. van Lange, A. W. Kruglanski, & E. T. Higgins (Eds.),

Handbook of theories of social psychology (Vol. 1, pp. 135–155). Sage. https://doi.org/10.4135/9781446249215.n8

Wells, L. M. (2024). *There are no deficits here: Disrupting anti-Blackness in education.* Corwin.

Wieseman, K. C., & Weinburgh, M. H. (2009). *Women's experiences in leadership in K-16 science education communities: Becoming and being.* Springer.

Yale School of Medicine. (2025). Yale Center for Emotional Intelligence. https://medicine.yale.edu/childstudy/services/community-and-schools-programs/center-for-emotional-intelligence/

Zachary, L. J., & Fain, L. Z. (2022). *The mentor's guide: Facilitating effective learning relationships.* Jossey-Bass.

Zehr, H. (2015). *The little book of restorative justice.* Good Books.

INDEX

ability *see* growth mindset
academic development *see* development
academic support 13
academic tenure *see* tenure
active listening 78
Adams, R.S. 5
affective language 101–102
affiliations 27, 114, 123
aggressions: macroaggressions 9; microaggressions 9
American Psychological Association 104
approachability *see* presence
Assessing Fit Checklist *see* fit
assimilation 120
Atkinson, J.W. 44
Attribution Theory of Motivation 44–45, 52, 142
autonomy *see* Self-Determination Theory

Bandura, A. 47
barriers 32, 90, 99, 116, 138–139, 143
Beaudoin, M. 99
Beech, N. 97
belonging 6, 10, 12–15, 21, 23–26, 48, 66, 80, 116, 118, 123
Beneficence and Nonmaleficence *see* ethics
body language *see* non-verbal cues

boundaries 69, 71, 75–77, 83, 85, 98–100, 104, 106, 108, 117, 121, 143
Brockbank, A. 97
Brown v. Board of Education 38
burnout 4, 10, 99, 143–144

challenge *see* facilitating learning
check-ins 62, 81
child development *see* development
Christianity *see* education history
circle processes 102–103
co-dependency 97, 100
collective labor 6, 117
colonization *see* education history
comfort zone *see* Zone of Proximal Development
communication 43, 61–64, 68, 81, 83, 90, 92–93, 101, 134
community *see* community of practice
community of practice 112–113
competence *see* Self-Determination Theory
competent leadership *see* leadership
conferences: professional 69, 79, 85, 89, 115, 123, 139; in restorative practices 103
confidentiality 75, 77
conflict *see* problem solving
conformity 119–120
connections *see* networks

contexts 4, 6, 10, 14, 20, 30, 43, 61, 66, 94, 107, 130–131, 134–135
Covey, S.R. 91
creativity 121–124
credit 104
critical consciousness 40
cultural competence 40
cultural responsiveness 36–37, 39–41, 43, 51–52, 90, 94, 96, 98, 105, 111, 129, 131, 137
culturally relevant pedagogy 40–42, 132
culture 6, 12, 37–40, 43, 52–53, 66, 95, 102, 114, 118, 120–121, 123–124, 137

dame schools *see* education history
deficit-minded approach 38–39, 89, 133
demographics *see* personal characteristics
dependency 97
development: academic 13, 50–51, 97–99, 118; child 21, 87; human 6, 8, 48, 87; identity 10, 14, 20, 23, 26–27, 31, 33, 50, 53, 137; social 10, 50, 76, 88, 140; professional 5, 76, 79, 89, 133; psychosocial 8, 10, 15, 40, 50, 80
developmental alliance 97
developmental networks *see* networks
developmental relationships 97, 111
developmental tie 117
diversity, in networks 22, 31, 66, 84, 117, 132
domain *see* community of practice
Dweck, C.S. 44, 50

education history: Christianity 37; colonial 37, 136; dame schools 38; enslaved 37; grammar schools 38; Indian boarding schools 38; Roman Catholic 37
effort 45, 48–51, 80, 83, 104
emotional intelligence: relationship management 94; self-awareness 94; self-management 94; social-awareness 94
empathic listening 32, 53, 71, 73–74, 90–94, 98, 106, 134
empathy 63, 90, 93–94, 98, 102, 104, 106, 143

enslaved *see* education history
Erikson, E.H. 14, 20
eSTEM Peer Mentoring 68, 83, 138–139
ethical mentoring *see* ethics
ethics: Beneficence and Nonmaleficence 104, 135; Fidelity and Responsibility 104–105; General Principles 104; Justice 104–105; Respect for People's Rights and Dignity 104–105
evaluation 78, 81, 93, 105, 114
expectation setting *see* relationship negotiation

facilitating learning: challenge 78, 80–82, 84, 88–90, 97; support 78–82; vision 78, 81
faculty facilitator, definition of 4
Fain, L.Z. 78, 82, 94–95
familial backgrounds *see* Institutional Departure Model
feedback 7, 16, 45, 50, 61, 63–64, 71, 80–81
Fidelity and Responsibility *see* ethics
fit 66–67, 73–74; Assessing Fit Checklist 74
fixed mindset 49, 50, 52

General Principles *see* ethics
generations 95
goals 10, 16, 26–27, 30, 32–33, 47–50, 52, 60–61, 63, 67–68, 71, 73–78, 80–84, 89–90, 92, 95, 98, 100–101, 108, 112–114, 116, 118, 120–122, 138
grammar schools *see* education history
growth mindset 36, 48–53, 80, 89–90, 98, 111, 135
guide on the side 60, 66, 68, 78, 81, 100

hierarchical boundary 97
hierarchical structure 96–97, 106
high-impact practice 6
historically Black colleges and universities (HBCUs) 26
human development *see* development

identity: identity compatibility 26, 31–33; Identity Theory 14–15, 20, 28, 32, 137
identity compatibility *see* identity

identity development *see* development
Identity Theory *see* identity
Indian boarding schools *see* education history
Indigenous 37–38, 102
Institutional Departure Model 12, 15
instrumental support 13
integration *see* Institutional Departure Model
integrity *see* ethics
interest, disciplinary 3, 6, 10, 11, 13–14, 33, 46, 66–67, 79, 84, 112–114, 119–120, 141
internships 69, 79, 89, 114
intersectionality 22, 138
introductions 27, 75
Irby, D.M. 26, 31

Justice *see* ethics

Keller, J.L. 5

Ladson-Billings, G. 39–41
leadership, transformative 98, 116, 137
legacy 36
LePera, N. 98
Let Them Theory 99, 144
listening, levels 91–92, 106–107

Maki, K. 99
Malott, C. 37
managing conflict *see* problem solving
marginalized populations 23–26, 32, 116, 121, 123, 133
match *see* fit
McGee, E.O. 5
mentee, definition of 10
mentee characteristics 64
mentor, definition of 10
mentor characteristics 61
mentoring: agreement 76–78, 122, 139; phases 68, 71–73, 75, 83–84, 89, 111; structures 3, 9, 16; training 10, 68, 78, 83–84, 129, 131, 138
mindfulness 99, 143
mindset *see* fixed mindset; growth mindset
mistakes 50–51, 67, 98–99, 117, 123
Model of Student Motivation and Persistence 48–49, 52
Mondisa, J. 5

motivation 3, 6, 13–15, 26, 43–53, 60, 62–63, 80, 111, 116, 134, 140–142, 145
Mullen, C.A. 5
mutualistic relationship *see* reciprocity

network diversity *see* networks
networking *see* networks
networks 11, 61, 69, 79, 85, 89, 112, 114–118, 123
non-verbal cues 43, 75
norms 53, 118, 120–121, 123–124
notes, meeting 75, 81

observation 16, 29, 43, 47, 87, 106
O'Sullivan, P.S. 26, 31

Packard, B.W. 5, 9
perceived worth *see* Model of Student Motivation and Persistence
performance accomplishment *see* Self-Efficacy Theory
persistence 5–6, 12–13, 15, 24, 26, 44–50, 80, 116, 140–141
personal attributes *see* Institutional Departure Model; personal characteristics
personal characteristics 74, 84, 95–96
personality type *see* Person-Environment Fit
Person-Environment Fit 119–120, 123–124
physiological response *see* Self-Efficacy Theory
poster 79, 123
potential *see* Zone of Proximal Development
power 8–9, 22, 51, 96–97, 99, 106–107, 135, 138, 142
power differential *see* power
power dynamics *see* power
practice *see* community of practice
praise 50, 51, 80
presence 42
prior experiences *see* Institutional Departure Model
problem solving 36, 66, 80–81, 88, 94, 103, 113, 130, 143
professional balance *see* boundaries
professional network 61, 69, 79, 89, 114–115
professional organizations *see* affiliations

psychological needs 46–47
psychosocial development *see* development
psychosocial support 5, 13, 96, 103
purpose 61, 67, 71, 75–76, 95

reciprocity 6, 8, 11, 52, 64, 105, 113, 116
reflection: self-reflection 27, 40, 43, 52, 64, 78, 81, 84, 105–108, 124, 132, 139; student reflection 64, 68, 82
relatedness *see* Self-Determination Theory
relationship building 71, 75, 84, 133
relationship constellations 117, 122–123
relationship management *see* emotional intelligence
relationship negotiation 74, 75, 84
relationship role shifts *see* relationship roles
relationship roles: faculty facilitator 9, 26, 30, 39–41, 65–66, 69, 71, 73, 78, 83–85, 87, 89, 95, 100–101, 104–106, 111, 114, 116, 118, 121, 123, 139; peer mentee 11, 63, 71, 82–84, 105, 111, 122, 139; peer mentor 60–61, 71, 82–84, 105, 111, 122, 139; shifts in 122–124
relationship structure *see* mentoring
relationship termination 71, 82, 84
relationship working 71, 78–81, 83–84, 89
research opportunities 69, 79, 85
respect 6, 11, 21, 33, 36, 41, 43, 52, 69–70, 90, 93, 100, 102–104, 106, 112, 122, 137
Respect for People's Rights and Dignity *see* ethics
responsibilities *see* relationship roles
restorative discipline 100–101; *see also* restorative practices
restorative justice practices *see* restorative practices
restorative practices 101–103, 106, 108, 142–143
retention 3, 6, 13, 24, 120, 141
Robbins, M. 99
role model 5, 30, 61, 63, 69, 83–84, 89, 114, 121
Rolfe, A. 89–91
Roman Catholic *see* education history

scaffold 4, 60, 81, 88, 134
scholarship *see* tenure
segregation 38
self-awareness *see* emotional intelligence
Self-Determination Theory: autonomy 46, 65, 69, 93, 97, 135; competence 13–14, 23, 46, 112; relatedness 46
self-efficacy *see* Self-Efficacy Theory
Self-Efficacy Theory: performance accomplishment 14–15; physiological response 14–15; self-efficacy 6, 13–15, 26, 48; social persuasion 14–15; vicarious experience 14–15
self-management *see* emotional intelligence
self-preservation *see* boundaries
sense of community 12, 15, 112–114, 116, 123
service *see* tenure
Situated Expectancy-Value Theory 44–46, 52, 140
skilled helper 6, 87–89
SMART goals 76
Social Cognitive Career Theory 13, 15, 116
social connectedness 46
social development *see* development
social persuasion *see* Self-Efficacy Theory
social-awareness *see* emotional intelligence
Social-Cognitive Theory 44, 47–48
Sociocultural Theory 87–88
sociopolitical consciousness *see* critical consciousness
stereotypes 8, 24–25, 43
strengths-based approach 89
student reflections *see* reflection
support *see* facilitating learning

talking piece *see* circle processes
teaching *see* tenure
tenure 105, 114–115
Theory of Psychosocial Development 14, 20
Tinto, V. 48–49, 52
training *see* mentoring
transformative leadership *see* leadership
trust 61–64, 71, 75, 80, 92–93, 105–107

Unconditional Positive Regard 93, 98, 106
underrepresented populations *see* marginalized populations

validation 80, 93
values 6, 27, 30, 36–37, 39, 41–42, 44, 46, 52–53, 69, 71, 105, 111, 120–121
vicarious experience *see* Self-Efficacy Theory
vision *see* facilitating learning

Weiner, B. 44–45
Wolfe, S. 99, 118
workbook 68, 132, 135, 139

Zachary, L.J. 78, 82, 94–95
Zone of Proximal Development 88

For Product Safety Concerns and Information please contact our EU representative GPSR@taylorandfrancis.com
Taylor & Francis Verlag GmbH, Kaufingerstraße 24, 80331 München, Germany

www.ingramcontent.com/pod-product-compliance
Lightning Source LLC
Chambersburg PA
CBHW061717300426
44115CB00014B/2727